True Bear Stories

Joaquin Miller

Alpha Editions

This edition published in 2024

ISBN : 9789362092250

Design and Setting By
Alpha Editions
www.alphaedis.com
Email - info@alphaedis.com

As per information held with us this book is in Public Domain.
This book is a reproduction of an important historical work. Alpha Editions uses the best technology to reproduce historical work in the same manner it was first published to preserve its original nature. Any marks or number seen are left intentionally to preserve its true form.

PREFACE.

My Bright Young Reader: I was once exactly your own age. Like all boys, I was, from the first, fond of bear stories, and above all, I did not like stories that seemed the least bit untrue. I always preferred a natural and reasonable story and one that would instruct as well as interest. This I think best for us all, and I have acted on this line in compiling these comparatively few bear stories from a long life of action in our mountains and up and down the continent.

As a rule, the modern bear is not a bloody, bad fellow, whatever he may have been in Bible days. You read, almost any circus season, about the killing of his keeper by a lion, a tiger, a panther, or even the dreary old elephant, but you never hear of a tame bear's hurting anybody.

I suppose you have been told, and believe, that bears will eat boys, good or bad, if they meet them in the woods. This is not true. On the contrary, there are several well-authenticated cases, in Germany mostly, where bears have taken lost children under their protection, one boy having been reared from the age of four to sixteen by a she bear without ever seeing the face of man.

I have known several persons to be maimed or killed in battles with bears, but in every case it was not the bear that began the fight, and in all my experience of about half a century I never knew a bear to eat human flesh, as does the tiger and like beasts.

Each branch of the bear family is represented here and each has its characteristics. By noting these as you go along you may learn something not set down in the schoolbooks. For the bear is a shy old hermit and is rarely encountered in his wild state by anyone save the hardy hunter, whose only interest in the event is to secure the skin and carcass.

Of course, now and then, a man of science meets a bear in the woods, but the meeting is of short duration. If the bear does not leave, the man of books does, and so we seldom get his photograph as he really appears in his wild state. The first and only bear I ever saw that seemed to be sitting for his photograph was the swamp, or "sloth," bear—Ursus Labiatus—found in the marshes at the mouth of the Mississippi River. You will read of an encounter with him further on.

I know very well that there exists a good deal of bad feeling between boys and bears, particularly on the part of boys. The trouble began, I suppose, about the time when that old she bear destroyed more than forty boys at a single meeting, for poking fun at a good old prophet. And we read that

David, when a boy, got very angry at a she bear and slew her single-handed and alone for interfering with his flock. So you see the feud between the boy and bear family is an old one indeed.

But I am bound to say that I have found much that is pathetic, and something that is almost half-human, in this poor, shaggy, shuffling hermit. He doesn't want much, only the wildest and most worthless parts of the mountains or marshes, where, if you will let him alone, he will let you alone, as a rule. Sometimes, out here in California, he loots a pig-pen, and now and then he gets among the bees. Only last week, a little black bear got his head fast in a bee-hive that had been improvised from a nail-keg, and the bee-farmer killed him with a pitchfork; but it is only when hungry and far from home that he seriously molests us.

The bear is a wise beast. This is, perhaps, because he never says anything. Next to the giraffe, which you may know never makes any noise or note whatever, notwithstanding the wonderful length of his throat, the bear is the most noiseless of beasts. With his nose to the ground all the time, standing up only now and then to pull a wild plum or pick a bunch of grapes, or knock a man down if he must, he seems to me like some weary old traveler that has missed the right road of life and doesn't quite know what to do with himself. Ah! if he would only lift up his nose and look about over this beautiful world, as the Indians say the grizzly bear was permitted to do before he disobeyed and got into trouble, an account of which you will find further on, why, the bear might be less a bear.

Stop here and reflect on how much there is in keeping your face well lifted. The pig with his snout to the ground will be forever a pig; the bear will be a bear to the end of his race, because he will not hold up his head in the world; but the horse—look at the horse! However, our business is with the bear now.

INTRODUCTORY NOTES.

The bear is the most human of all the beasts. He is not the most man-like in anatomy, nor the nearest in the line of evolution. The likeness is rather in his temper and way of doing things and in the vicissitudes of his life. He is a savage, of course, but most men are that—wild members of a wild fauna—and, like wild men, the bear is a clumsy, good-natured blunderer, eating with his fingers in default of a knife, and preferring any day a mouthful of berries to the excitement of a fight.

In this book Joaquin Miller has tried to show us the bear as he is, not the traditional bear of the story-books. In season and out of season, the bear has been represented always the same bear, "as much alike as so many English noblemen in evening dress," and always as a bloody bear.

Mr. Miller insists that there are bears and bears, as unlike one another in nature and action as so many horses, hogs or goats. This mon—*bears are never cruel.* They are generally full of homely, careless kindness, and are very fond of music as well as of honey, blackberries, nuts, fish and other delicacies of the savage feast.

The matter of season affects a bear's temper and looks as the time of the day affects those of a man.

He goes to bed in the fall, when the fish and berry season is over, fat and happy, with no fight in him. He comes out in spring, just as good-natured, if not so fat. But the hot sun melts him down. His hungry hunt for roots, bugs, ants and small game makes him lean and cross. His claws grow long, his hair is unkempt and he is soon a shaggy ghost of himself, looking "like a second-hand sofa with the stuffing coming out," and in this out-at-elbows condition he loses his own self-respect.

Mr. Miller has strenuously insisted that bears of the United States are of more than one or two species. In this he has the unqualified support of the latest scientific investigations. Not long ago naturalists were disposed to recognize but three kinds of bear in North America. These are the polar bear, the black bear, and the grizzly bear, and even the grizzly was thought doubtful, a slight variation of the bear of Europe.

But the careful study of bears' skulls has changed all that, and our highest authority on bears, Dr. C. Hart Merriam of the Department of Agriculture, now recognizes not less than ten species of bear in the limits of the United States and Alaska.

In his latest paper (1896), a "Preliminary Synopsis of the American Bears," Dr. Merriam groups these animals as follows:

I. POLAR BEARS.

1. POLAR BEAR: *Thalarctos maritimus* Linnaeus. Found on all Arctic shores.

II. BLACK BEARS.

2. COMMON BLACK BEAR (sometimes brown or cinnamon): *Ursus americanus* Pallas. Found throughout the United States.

3. YELLOW BEAR (sometimes black or brown): *Ursus luteolus* Griffith. Swamps of Louisiana and Texas.

4. EVERGLADE BEAR: *Ursus floridanus* Merriam. Everglades of Florida.

5. GLACIER BEAR: *Ursus emmonsi* Dall. About Mount St. Elias.

III. GRIZZLY BEARS.

6. THE GRIZZLY BEAR: *Ursus horribilis* Ord. Found in the western parts of North America.

Under this species are four varieties: the original *horribilis*, or Rocky Mountain grizzly, from Montana to the Great Basin of Utah; the variety *californicus* Merriam, the California grizzly, from the Sierra Nevada; variety *horriaeus* Baird, the Sonora grizzly, from Arizona and the South; and variety *alascensis* Merriam, the Alaska grizzly, from Alaska.

7. THE BARREN GROUND BEAR: *Ursus richardsoni* Mayne Reid. A kind of grizzly found about Hudson Bay.

IV. GREAT BROWN BEARS.

8. THE YAKUTAT BEAR: *Ursus dalli* Merriam. From about Mount St. Elias.

9. THE SITKA BEAR: *Ursus sitkensis* Merriam. From about Sitka.

10. THE KADIAK BEAR: *Ursus middendorfi* Merriam. From Kadiak and the Peninsula of Alaska.

These three bears are even larger than the grizzly, and the Kadiak Bear is the largest of all the land bears of the world. It prowls about over the moss of the mountains, feeding on berries and fish.

The sea-bear, *Callorhinus ursinus*, which we call the fur seal, is also a cousin of the bear, having much in common with its bear ancestors of long ago, but neither that nor its relations, the sea-lion and the walrus, are exactly bears to-day.

Of all the real bears, Mr. Miller treats of five in the pages of this little book. All the straight "bear stories" relate to *Ursus americanus*, as most bear stories in our country do. The grizzly stories treat of *Ursus horribilis californicus*. The lean bear of the Louisiana swamps is *Ursus luteolus*, and the Polar Bear is *Thalarctos maritimus*. The author of the book has tried without intrusion of technicalities to bring the distinctive features of the different bears before the reader and to instruct as well as to interest children and children's parents in the simple realities of bear life.

David Starr Jordan.

Leland Stanford, Jr., University.

I.

A BEAR ON FIRE.

It is now more than a quarter of a century since I saw the woods of Mount Shasta in flames, and beasts of all sorts, even serpents, crowded together; but I can never forget, never!

It looked as if we would have a cloudburst that fearful morning. We three were making our way by slow marches from Soda Springs across the south base of Mount Shasta to the Modoc lava beds—two English artists and myself. We had saddle horses, or, rather, two saddle horses and a mule, for our own use. Six Indians, with broad leather or elkskin straps across their foreheads, had been chartered to carry the kits and traps. They were men of means and leisure, these artists, and were making the trip for the fish, game, scenery and excitement and everything, in fact, that was in the adventure. I was merely their hired guide.

This second morning out, the Indians—poor slaves, perhaps, from the first, certainly not warriors with any spirit in them—began to sulk. They had risen early and kept hovering together and talking, or, rather, making signs in the gloomiest sort of fashion. We had hard work to get them to do anything at all, and even after breakfast was ready they packed up without tasting food.

The air was ugly, for that region—hot, heavy, and without light or life. It was what in some parts of South America they call "earthquake weather." Even the horses sulked as we mounted; but the mule shot ahead through the brush at once, and this induced the ponies to follow.

The Englishmen thought the Indians and horses were only tired from the day before, but we soon found the whole force plowing ahead through the dense brush and over fallen timber on a double quick.

Then we heard low, heavy thunder in the heavens. Were they running away from a thunder-storm? The English artists, who had been doing India and had come to love the indolent patience and obedience of the black people, tried to call a halt. No use. I shouted to the Indians in their own tongue. "Tokau! Ki-sa! Kiu!" (Hasten! Quick! Quick!) was all the answer I could get from the red, hot face that was thrown for a moment back over the load and shoulder. So we shot forward. In fact, the horses now refused all regard for the bit, and made their own way through the brush with wondrous skill and speed.

We were flying from fire, not flood! Pitiful what a few years of neglect will do toward destroying a forest! When a lad I had galloped my horse in security and comfort all through this region. It was like a park then. Now it was a dense tangle of undergrowth and a mass of fallen timber. What a feast for flames! In one of the very old books on America in the British Museum—possibly the very oldest on the subject—the author tells of the park-like appearance of the American forests. He tells his English friends back at home that it is most comfortable to ride to the hounds, "since the Indian squats (squaws) do set fire to the brush and leaves every spring," etc.

But the "squats" had long since disappeared from the forests of Mount Shasta; and here we were tumbling over and tearing through ten years' or more of accumulation of logs, brush, leaves, weeds and grass that lay waiting for a sea of fire to roll over all like a mass of lava.

And now the wind blew past and over us. Bits of white ashes sifted down like snow. Surely the sea of fire was coming, coming right on after us! Still there was no sign, save this little sift of ashes, no sound; nothing at all except the trained sense of the Indians and the terror of the "cattle" (this is what the Englishmen called our horses) to give us warning.

In a short time we struck an arroyo, or canyon, that was nearly free from brush and led steeply down to the cool, deep waters of the McCloud River. Here we found the Indians had thrown their loads and themselves on the ground.

They got up in sulky silence, and, stripping our horses, turned them loose; and then, taking our saddles, they led us hastily up out of the narrow mouth of the arroyo under a little steep stone bluff.

They did not say a word or make any sign, and we were all too breathless and bewildered to either question or protest. The sky was black, and thunder made the woods tremble. We were hardly done wiping the blood and perspiration from our torn hands and faces where we sat when the mule jerked up his head, sniffed, snorted and then plunged headlong into the river and struck out for the deep forest on the farther bank, followed by the ponies.

The mule is the most traduced of all animals. A single mule has more sense than a whole stableful of horses. You can handle a mule easily if the barn is burning; he keeps his head; but a horse becomes insane. He will rush right into the fire, if allowed to, and you can only handle him, and that with difficulty if he sniffs the fire, by blindfolding him. Trust a mule in case of peril or a panic long before a horse. The brother of Solomon and willful son of David surely had some of the great temple-builder's wisdom and discernment, for we read that he rode a mule. True, he lost his head and got hung up by the hair, but that is nothing against the mule.

As we turned our eyes from seeing the animals safely over, right there by us and a little behind us, through the willows of the canyon and over the edge of the water, we saw peering and pointing toward the other side dozens of long black and brown outreaching noses. Elk!

They had come noiselessly, they stood motionless. They did not look back or aside, only straight ahead. We could almost have touched the nearest one. They were large and fat, almost as fat as cows; certainly larger than the ordinary Jersey. The peculiar thing about them was the way, the level way, in which they held their small, long heads—straight out; the huge horns of the males lying far back on their shoulders. And then for the first time I could make out what these horns are for—to part the brush with as they lead through the thicket, and thus save their coarse coats of hair, which is very rotten, and could be torn off in a little time if not thus protected. They are never used to fight with, never; the elk uses only his feet. If on the defense, however, the male elk will throw his nose close to the ground and receive the enemy on his horns.

Suddenly and all together, and perhaps they had only paused a second, they moved on into the water, led by a bull with a head of horns like a rocking-chair. And his rocking-chair rocked his head under water much of the time. The cold, swift water soon broke the line, only the leader making the bank directly before us, while the others drifted far down and out of sight.

Our artists, meantime, had dug up pencil and pad and begun work. But an Indian jerked the saddles, on which the Englishmen sat, aside, and the work was stopped. Everything was now packed up close under the steep little ledge of rocks. An avalanche of smaller wild animals, mostly deer, was upon us. Many of these had their tongues hanging from their half-opened mouths. They did not attempt to drink, as you would suppose, but slid into the water silently almost as soon as they came. Surely they must have seen us, but certainly they took no notice of us. And such order! No crushing or crowding, as you see cattle in corrals, aye, as you see people sometimes in the cars.

And now came a torrent of little creeping things: rabbits, rats, squirrels! None of these smaller creatures attempted to cross, but crept along in the willows and brush close to the water.

They loaded down the willows till they bent into the water, and the terrified little creatures floated away without the least bit of noise or confusion. And still the black skies were filled with the solemn boom of thunder. In fact, we had not yet heard any noise of any sort except thunder, not even our own voices. There was something more eloquent in the air now, something more terrible than man or beast, and all things were awed into silence—a profound silence.

And all this time countless creatures, little creatures and big, were crowding the bank on our side or swimming across or floating down, down, down the swift, woodhung waters. Suddenly the stolid leader of the Indians threw his two naked arms in the air and let them fall, limp and helpless at his side; then he pointed out into the stream, for there embers and living and dead beasts began to drift and sweep down the swift waters from above. The Indians now gathered up the packs and saddles and made a barricade above, for it was clear that many a living thing would now be borne down upon us.

The two Englishmen looked one another in the face long and thoughtfully, pulling their feet under them to keep from being trodden on. Then, after another avalanche of creatures of all sorts and sizes, a sort of Noah's ark this time, one of them said to the other:

"Beastly, you know!"

"Awful beastly, don't you know!"

As they were talking entirely to themselves and in their own language, I did not trouble myself to call their attention to an enormous yellow rattlesnake which had suddenly and noiselessly slid down, over the steep little bluff of rocks behind us, into our midst.

But now note this fact—every man there, red or white, saw or felt that huge and noiseless monster the very second she slid among us. For as I looked, even as I first looked, and then turned to see what the others would say or do, they were all looking at the glittering eyes set in that coffin-like head.

The Indians did not move back or seem nearly so much frightened as when they saw the drift of embers and dead beasts in the river before them; but the florid Englishmen turned white! They resolutely arose, thrust their hands in their pockets and stood leaning their backs hard against the steep bluff. Then another snake, long, black and beautiful, swept his supple neck down between them and thrust his red tongue forth—as if a bit of the flames had already reached us.

Fortunately, this particular "wisest of all the beasts of the field," was not disposed to tarry. In another second he had swung to the ground and was making a thousand graceful curves in the swift water for the further bank.

The world, even the world of books, seems to know nothing at all about the wonderful snakes that live in the woods. The woods rattlesnake is as large as at least twenty ordinary rattlesnakes; and Indians say it is entirely harmless. The enormous black snake, I know, is entirely without venom. In all my life, spent mostly in the camp, I have seen only three of those monstrous yellow woods rattlesnakes; one in Indiana, one in Oregon and the other on this occasion here on the banks of the McCloud. Such bright eyes! It was hard to stop looking at them.

Meantime a good many bears had come and gone. The bear is a good swimmer, and takes to the water without fear. He is, in truth, quite a fisherman; so much of a fisherman, in fact, that in salmon season here his flesh is unfit for food. The pitiful part of it all was to see such little creatures as could not swim clinging all up and down and not daring to take to the water.

Unlike his domesticated brother, we saw several wild-cats take to the water promptly. The wild-cat, you must know, has no tail to speak of. But the panther and Californian lion are well equipped in this respect and abhor the water.

I constantly kept an eye over my shoulder at the ledge or little bluff of rocks, expecting to see a whole row of lions and panthers sitting there, almost "cheek by jowl" with my English friends, at any moment. But strangely enough, we saw neither panther nor lion; nor did we see a single grizzly among all the bears that came that way.

We now noticed that one of the Indians had become fascinated or charmed by looking too intently at the enormous serpent in our midst. The snake's huge, coffin-shaped head, as big as your open palm, was slowly swaying from side to side. The Indian's head was doing the same, and their eyes were drawing closer and closer together. Whatever there may be in the Bible story of Eve and the serpent, whether a figure or a fact, who shall say?—but it is certainly, in some sense, true.

An Indian will not kill a rattlesnake. But to break the charm, in this case, they caught their companion by the shoulders and forced him back flat on the ground. And there he lay, crying like a child, the first and only Indian I ever saw cry. And then suddenly boom! boom! boom! as if heaven burst. It began to rain in torrents.

And just then, as we began to breathe freely and feel safe, there came a crash and bump and bang above our heads, and high over our heads from off the ledge behind us! Over our heads like a rocket, in an instant and clear into the water, leaped a huge black bear, a ball of fire! his fat sides in flame. He sank out of sight but soon came up, spun around like a top, dived again, then again spun around. But he got across, I am glad to say. And this always pleases my little girl, Juanita. He sat there on the bank looking back at us quite a time. Finally he washed his face, like a cat, then quietly went away. The rattlesnake was the last to cross.

Into the water leaped a black bear.

The beautiful yellow beast was not at all disconcerted, but with the serenest dignity lifted her yellow folds, coiled and uncoiled slowly, curved high in the air, arched her glittering neck of gold, widened her body till broad as your two hands, and so slid away over the water to the other side through the wild white rain. The cloudburst put out the fire instantly, showing that, though animals have superhuman foresight, they don't know everything before the time.

"Beastly! I didn't get a blawsted sketch, you know."

"Awful beastly! Neither did I, don't you know."

And that was all my English friends said. The Indians made their moaning and whimpering friend who had been overcome by the snake pull himself together and they swam across and gathered up the "cattle."

Some men say a bear cannot leap; but I say there are times when a bear can leap like a tiger. This was one of the times.

II.

MUSIC-LOVING BEARS.

No, don't despise the bear, either in his life or his death. He is a kingly fellow, every inch a king; a curious, monkish, music-loving, roving Robin Hood of his somber woods—a silent monk, who knows a great deal more than he tells. And please don't go to look at him and sit in judgment on him behind the bars. Put yourself in his place and see how much of manhood or kinghood would be left in you with a muzzle on your mouth, and only enough liberty left to push your nose between two rusty bars and catch the peanut which the good little boy has found to be a bad one and so generously tosses it to the bear.

Of course, the little boy, remembering the experience of about forty other little boys in connection with the late baldheaded Elijah, has a prejudice against the bear family, but why the full-grown man should so continually persist in caging this shaggy-coated, dignified, kingly and ancient brother of his, I cannot see, unless it is that he knows almost nothing at all of his better nature, his shy, innocent love of a joke, his partiality for music and his imperial disdain of death. And so, with a desire that man may know a little more about this storied and classic creature which, with noiseless and stately tread, has come down to us out of the past, and is as quietly passing away from the face of the earth, these fragmentary facts are set down. But first as to his love of music. A bear loves music better than he loves honey, and that is saying that he loves music better than he loves his life.

We were going to mill, father and I, and Lyte Howard, in Oregon, about forty years ago, with ox-teams, a dozen or two bags of wheat, threshed with a flail and winnowed with a wagon cover, and were camped for the night by the Calipoola River; for it took two days to reach the mill. Lyte got out his fiddle, keeping his gun, of course, close at hand. Pretty soon the oxen came down, came very close, so close that they almost put their cold, moist noses against the backs of our necks as we sat there on the ox-yokes or reclined in our blankets, around the crackling pine-log fire and listened to the wild, sweet strains that swept up and down and up till the very tree tops seemed to dance and quiver with delight.

Then suddenly father seemed to feel the presence of something or somebody strange, and I felt it, too. But the fiddler felt, heard, saw nothing but the divine, wild melody that made the very pine trees dance and quiver to their tips. Oh, for the pure, wild, sweet, plaintive music once more! the music of "Money Musk," "Zip Coon," "Ol' Dan Tucker" and all the other dear old airs that once made a thousand happy feet keep time on the puncheon floors from Hudson's bank to the Oregon. But they are no more, now. They have passed away forever with the Indian, the pioneer, and the music-loving bear. It is strange how a man—I mean the natural man—will feel a presence long before he hears it or sees it. You can always feel the approach of a—but I forget. You are of another generation, a generation that only reads, takes thought at second hand only, if at all, and you would not understand; so let us get forward and not waste time in explaining the unexplainable to you.

Father got up, turned about, put me behind him like, as an animal will its young, and peered back and down through the dense tangle of the deep river bank between two of the huge oxen which had crossed the plains with us to the water's edge; then he reached around and drew me to him with his left hand, pointing between the oxen sharp down the bank with his right forefinger.

A bear! two bears! and another coming; one already more than half way across on the great, mossy log that lay above the deep, sweeping waters of the Calipoola; and Lyte kept on, and the wild, sweet music leaped up and swept through the delighted and dancing boughs above. Then father reached back to the fire and thrust a long, burning bough deeper into the dying embers and the glittering sparks leaped and laughed and danced and swept out and up and up as if to companion with the stars. Then Lyte knew. He did not hear, he did not see, he only felt; but the fiddle forsook his fingers and his chin in a second, and his gun was to his face with the muzzle thrust down between the oxen. And then my father's gentle hand reached out, lay on that long, black, Kentucky rifle barrel, and it dropped down, slept once more at the fiddler's side, and again the melodies; and the very stars came down, believe me, to listen, for they never seemed so big and so close by before. The bears sat down on their haunches at last, and one of them kept opening his mouth and putting out his red tongue, as if he really wanted to taste the music. Every now and then one of them would lift up a paw and gently tap the ground, as if to keep time with the music. And both my papa and Lyte said next day that those bears really wanted to dance.

And that is all there is to say about that, except that my father was the gentlest gentleman I ever knew and his influence must have been boundless; for who ever before heard of any hunter laying down his rifle with a family of fat black bears holding the little snow-white cross on their breasts almost within reach of its muzzle?

The moon came up by and by, and the chin of the weary fiddler sank lower and lower, till all was still. The oxen lay down and ruminated, with their noses nearly against us. Then the coal-black bears melted away before the milk-white moon, and we slept there, with the sweet breath of the cattle, like incense, upon us.

But how does a bear die? Ah, I had forgotten. I must tell you of death, then. Well, we have different kinds of bears. I know little of the Polar bear, and so say nothing positively of him. I am told, however, that there is not, considering his size, much snap or grit about him; but as for the others, I am free to say that they live and die like gentlemen.

I shall find time, as we go forward, to set down many incidents out of my own experience to prove that the bear is often a humorist, and never by any means a bad fellow.

Judge Highton, odd as it may seem, has left the San Francisco bar for the "bar" of Mount Shasta every season for more than a quarter of a century, and he probably knows more about bears than any other eminently learned man in the world, and Henry Highton will tell you that the bear is a good fellow at home, good all through, a brave, modest, sober old monk.

> A monkish Robin Hood
>
> In his good green wood.

III.

MY FIRST GRIZZLY.

One of Fremont's men, Mountain Joe, had taken a fancy to me down in Oregon, and finally, to put three volumes in three lines, I turned up as partner in his Soda Springs ranch on the Sacramento, where the famous Shasta-water is now bottled, I believe. Then the Indians broke out, burned us up and we followed and fought them in Castle rocks, and I was shot down. Then my father came on to watch by my side, where I lay, under protection of soldiers, at the mouth of Shot Creek canyon.

As the manzanita berries began to turn the mountain sides red and the brown pine quills to sift down their perfumed carpets at our feet, I began to feel some strength and wanted to fight, but I had had enough of Indians. I wanted to fight grizzly bears this time. The fact is, they used to leave tracks in the pack trail every night, and right close about the camp, too, as big as the head of a barrel.

Now father was well up in woodcraft, no man better, but he never fired a gun. Never, in his seventy years of life among savages, did that gentle Quaker, school-master, magistrate and Christian ever fire a gun. But he always allowed me to have my own way as a hunter, and now that I was getting well of my wound he was so glad and grateful that he willingly joined in with the soldiers to help me kill one of these huge bears that had made the big tracks.

Do you know why a beast, a bear of all beasts, is so very much afraid of fire? Well, in the first place, as said before, a bear is a gentleman, in dress as well as address, and so likes a decent coat. If a bear should get his coat singed he would hide away from sight of both man and beast for half a year. But back of his pride is the fact that a fat bear will burn like a candle; the fire will not stop with the destruction of his coat. And so, mean as it was, in the olden days, when bears were as common in California as cows are now, men used to take advantage of this fear and kindle pine-quill fires in and around his haunts in the head of canyons to drive him out and down and into ambush.

Read two or three chapters here between the lines—lots of plans, preparations, diagrams. I was to hide near camp and wait—to place the crescent of pine-quill fires and all that. Then at twilight they all went out and away on the mountain sides around the head of the canyon, and I hid behind a big rock near by the extinguished camp-fire, with my old muzzle-loading Kentucky rifle, lifting my eyes away up and around to the head of the Manzanita canyon looking for the fires. A light! One, two, three, ten! A sudden crescent of forked flames, and all the fight and impetuosity of a boy of only a dozen years was uppermost, and I wanted a bear!

All alone I waited; got hot, cold, thirsty, cross as a bear and so sick of sitting there that I was about to go to my blankets, for the flames had almost died out on the hills, leaving only a circle of little dots and dying embers, like a fading diadem on the mighty lifted brow of the glorious Manzanita mountain. And now the new moon came, went softly and sweetly by, like a shy, sweet maiden, hiding down, down out of sight.

Crash! His head was thrown back, not over his shoulder, as you may read but never see, but down by his left foot, as he looked around and back up the brown mountain side. He had stumbled, or rather, he had stepped on himself, for a bear gets down hill sadly. If a bear ever gets after you, you had better go hill and go down hill fast. It will make him mad, but that is not your affair. I never saw a bear go down hill in a good humor. What nature meant by making a bear so short in the arms I don't know. Indians say he was first a man and walked upright with a club on his shoulder, but sinned and fell. As evidence of this, they show that he can still stand up and fight with his fists when hard pressed, but more of this later on.

This huge brute before me looked almost white in the tawny twilight as he stumbled down through the steep tangle of chaparral into the opening on the stony bar of the river.

He had evidently been terribly tangled up and disgusted while in the bush and jungle, and now, well out of it, with the foamy, rumbling, roaring Sacramento River only a few rods beyond him, into which he could plunge with his glossy coat, he seemed to want to turn about and shake his huge fists at the crescent of fire in the pine-quills that had driven him down the mountain. He threw his enormous bulk back on his haunches and rose up, and rose up, and rose up! Oh, the majesty of this king of our continent, as he seemed to still keep rising! Then he turned slowly around on his great hinder feet to look back; he pushed his nose away out, then drew it back, twisted his short, thick neck, like that of a beer-drinking German, and then for a final observation he tiptoed up, threw his high head still higher in the air and wiggled it about and sniffed and sniffed and—bang!

I shot at him from ambush, with his back toward me, shot at his back! For shame! Henry Highton would not have done that; nor, indeed, would I or any other real sportsman do such a thing now; but I must plead the "Baby Act," and all the facts, and also my sincere penitence, and proceed.

The noble brute did not fall, but let himself down with dignity and came slowly forward. Hugely, ponderously, solemnly, he was coming. And right here, if I should set down what I thought about—where father was, the soldiers, anybody, everybody else, whether I had best just fall on my face and "play possum" and put in a little prayer or two on the side, like—well, I was going on to say that if I should write all that flashed and surged through my mind in the next three seconds, you would be very tired. I was certain I had not hit the bear at all. As a rule, you can always see the "fur fly," as hunters put it; only it is not fur, but dust, that flies.

But this bear was very fat and hot, and so there could have been no dust to fly. After shuffling a few steps forward and straight for the river, he suddenly surged up again, looked all about, just as before, then turned his face to the river and me, the tallest bear that ever tiptoed up and up and up in the Sierras. One, two, three steps—on came the bear! and my gun empty! Then he fell, all at once and all in a heap. No noise, no moaning or groaning at all, no clutching at the ground, as men have seen Indians and even white men do; as if they would hold the earth from passing away—nothing of that sort. He lay quite still, head down hill, on his left side, gave just one short, quick breath, and then, pulling up his great right paw, he pushed his nose and eyes under it, as if to shut out the light forever, or, maybe, to muffle up his face as when "great Cæsar fell."

And that was all. I had killed a grizzly bear; nearly as big as the biggest ox.

He threw his enormous bulk back on his haunches, and rose up.

IV.

TWIN BABIES.

These twin babies were black. They were black as coal. Indeed, they were blacker than coal, for they glistened in their oily blackness. They were young baby bears; and so exactly alike that no one could, in any way, tell the one from the other. And they were orphans. They had been found at the foot of a small cedar tree on the banks of the Sacramento River, near the now famous Soda Springs, found by a tow-headed boy who was very fond of bears and hunting.

But at the time the twin babies were found Soda Springs was only a wild camp, or way station, on the one and only trail that wound through the woods and up and down mountains for hundreds of miles, connecting the gold fields of California with the pastoral settlements away to the north in Oregon. But a railroad has now taken the place of that tortuous old packtrail, and you can whisk through these wild and woody mountains, and away on down through Oregon and up through Washington, Montana, Dakota, Minnesota, Wisconsin and on to Chicago without even once getting out of your car, if you like. Yet such a persistent ride is not probable, for fish, pheasants, deer, elk, and bear still abound here in their ancient haunts, and the temptation to get out and fish or hunt is too great to be resisted.

This place where the baby bears were found was first owned by three men or, rather, by two men and a boy. One of the men was known as Mountain Joe. He had once been a guide in the service of General Fremont, but he was now a drunken fellow and spent most of his time at the trading post, twenty miles down the river. He is now an old man, almost blind, and lives in Oregon City, on a pension received as a soldier of the Mexican war. The other man's name was Sil Reese. He, also, is living and famously rich—as rich as he is stingy, and that is saying that he is very rich indeed.

The boy preferred the trees to the house, partly because it was more pleasant and partly because Sil Reese, who had a large nose and used it to talk with constantly, kept grumbling because the boy, who had been wounded in defending the ranch, was not able to work—wash the dishes, make fires and so on, and help in a general and particular way about the so-called "Soda Spring Hotel." This Sil Reese was certainly a mean man, as has, perhaps, been set down in this sketch before.

The baby bears were found asleep, and alone. How they came to be there, and, above all, how they came to be left long enough alone by their mother for a feeble boy to rush forward at sight of them, catch them up in his arms and escape with them, will always be a wonder. But this one thing is certain, you had about as well take up two rattlesnakes in your arms as two baby bears, and hope to get off unharmed, if the mother of the young bears is within a mile of you. This boy, however, had not yet learned caution, and he probably was not born with much fear in his make-up. And then he was so lonesome, and this man Reese was so cruel and so cross, with his big nose like a sounding fog-horn, that the boy was glad to get even a bear to love and play with.

They, so far from being frightened or cross, began to root around under his arms and against his breast, like little pigs, for something to eat. Possibly their mother had been killed by hunters, for they were nearly famished. When he got them home, how they did eat! This also made Sil Reese mad. For, although the boy, wounded as he was, managed to shoot down a deer not too far from the house almost every day, and so kept the "hotel" in meat, still it made Reese miserable and envious to see the boy so happy with his sable and woolly little friends. Reese was simply mean!

Before a month the little black boys began to walk erect, carry stick muskets, wear paper caps, and march up and down before the door of the big log "hotel" like soldiers.

But the cutest trick they learned was that of waiting on the table. With little round caps and short white aprons, the little black boys would stand behind the long bench on which the guests sat at the pine board table and pretend to take orders with all the precision and solemnity of Southern negroes.

Of course, it is to be confessed that they often dropped things, especially if the least bit hot; but remember we had only tin plates and tin or iron dishes of all sorts, so that little damage was done if a dish did happen to fall and rattle down on the earthen floor.

Men came from far and near and often lingered all day to see these cunning and intelligent creatures perform.

About this time Mountain Joe fought a duel with another mountaineer down at the trading post, and this duel, a bloodless and foolish affair, was all the talk. Why not have the little black fellows fight a duel also? They were surely civilized enough to fight now!

And so, with a very few days' training, they fought a duel exactly like the one in which poor, drunken old Mountain Joe was engaged; even to the detail of one of them suddenly dropping his stick gun and running away and falling headlong in a prospect hole.

When Joe came home and saw this duel and saw what a fool he had made of himself, he at first was furiously angry. But it made him sober, and he kept sober for half a year. Meantime Reese was mad as ever, more mad, in fact, than ever before. For he could not endure to see the boy have any friends of any kind. Above all, he did not want Mountain Joe to stay at home or keep sober. He wanted to handle all the money and answer no questions. A drunken man and a boy that he could bully suited him best. Ah, but this man Reese was a mean fellow, as has been said a time or two before.

As winter came on the two blacks were fat as pigs and fully half-grown. Their appetites increased daily, and so did the anger and envy of Mr. Sil Reese.

"They'll eat us out o' house and hum," said the big, towering nose one day, as the snow began to descend and close up the pack trails. And then the stingy man proposed that the blacks should be made to hibernate, as others of their kind. There was a big, hollow log that had been sawed off in joints to make bee gums; and the stingy man insisted that they should be put in there with a tight head, and a pack of hay for a bed, and nailed up till spring to save provisions.

Soon there was an Indian outbreak. Some one from the ranch, or "hotel," must go with the company of volunteers that was forming down at the post for a winter campaign. Of course Reese would not go. He wanted Mountain Joe to go and get killed. But Joe was sober now and he wanted to stay and watch Reese.

And that is how it came about that the two black babies were tumbled headlong into a big, black bee gum, or short, hollow log, on a heap of hay, and nailed up for the winter. The boy had to go to the war.

It was late in the spring when the boy, having neglected to get himself killed, to the great disgust of Mr. Sil Reese, rode down and went straight up to the big black bee gum in the back yard. He put his ear to a knothole. Not a sound. He tethered his mule, came back and tried to shake the short, hollow log. Not a sound or sign or movement of any kind. Then he kicked the big black gum with all his might. Nothing. Rushing to the wood-pile, he caught up an ax and in a moment had the whole end of the big gum caved in, and, to his infinite delight, out rolled the twins!

But they were merely the ghosts of themselves. They had been kept in a month or more too long, and were now so weak and so lean that they could hardly stand on their feet.

"Kill 'em and put 'em out o' misery," said Reese, for run from him they really could not, and he came forward and kicked one of them flat down on its face as it was trying hard to stand on its four feet.

The boy had grown some; besides, he was just from the war and was now strong and well. He rushed up in front of Reese, and he must have looked unfriendly, for Sil Reese tried to smile, and then at the same time he turned hastily to go into the house. And when he got fairly turned around, the boy kicked him precisely where he had kicked the bear. And he kicked him hard, so hard that he pitched forward on his face just as the bear had done. He got up quickly, but he did not look back. He seemed to have something to do in the house.

In a month the babies, big babies now, were sleek and fat. It is amazing how these creatures will eat after a short nap of a few months, like that. And their cunning tricks, now! And their kindness to their master! Ah! their glossy black coats and their brilliant black eyes!

And now three men came. Two of these men were Italians from San Francisco. The third man was also from that city, but he had an amazing big nose and refused to eat bear meat. He thought it was pork.

They took tremendous interest in the big black twins, and stayed all night and till late next day, seeing them perform.

"Seventy-five dollars," said one big nose to the other big nose, back in a corner where they thought the boy did not hear.

"One hundred and fifty. You see, I'll have to give my friends fifty each. Yes, it's true I've took care of 'em all winter, but I ain't mean, and I'll only keep fifty of it."

The boy, bursting with indignation, ran to Mountain Joe with what he had heard. But poor Joe had been sober for a long time, and his eyes fairly danced in delight at having $50 in his own hand and right to spend it down at the post.

And so the two Italians muzzled the big, pretty pets and led them kindly down the trail toward the city, where they were to perform in the streets, the man with the big nose following after the twins on a big white mule.

And what became of the big black twin babies? They are still performing, seem content and happy, sometimes in a circus, sometimes in a garden, sometimes in the street. They are great favorites and have never done harm to anyone.

And what became of Sil Reese? Well, as said before, he still lives, is very rich and very miserable. He met the boy—the boy that was—on the street the other day and wanted to talk of old times. He told the boy he ought to write something about the old times and put him, Sil Reese, in it. He said, with that same old sounding nose and sickening smile, that he wanted the boy to be sure and put his, Sil Reese's name, in it so that he could show it to his friends. And the boy has done so.

The boy? You want to know what the boy is doing? Well, in about a second he will be signing his autograph to the bottom of this story about his twin babies.

V.

IN SWIMMING WITH A BEAR.

What made these ugly rows of scars on my left hand?

Well, it might have been buckshot; only it wasn't. Besides, buckshot would be scattered about, "sort of promiscuous like," as backwoodsmen say. But these ugly little holes are all in a row, or rather in two rows. Now a wolf might have made these holes with his fine white teeth, or a bear might have done it with his dingy and ugly teeth, long ago. I must here tell you that the teeth of a bear are not nearly so fine as the teeth of a wolf. And the teeth of a lion are the ugliest of them all. They are often broken and bent; and they are always of a dim yellow color. It is from this yellow hue of the lion's teeth that we have the name of one of the most famous early flowers of May: dent de lion, tooth of the lion; dandelion. Get down your botany, now, find the Anglo-Asian name of the flower, and fix this fact on your mind before you read further.

I know of three men, all old men now, who have their left hands all covered with scars. One is due to the wolf; the others owe their scars to the red mouths of black bears.

You see, in the old days, out here in California, when the Sierras were full of bold young fellows hunting for gold, quite a number of them had hand-to-hand battles with bears. For when we came out here "the woods were full of 'em."

Of course, the first thing a man does when he finds himself face to face with a bear that won't run and he has no gun—and that is always the time when he finds a bear—why, he runs, himself; that is, if the bear will let him.

But it is generally a good deal like the old Crusader who "caught a Tartar" long ago, when on his way to capture Jerusalem, with Peter the Hermit.

"Come on!" cried Peter to the helmeted and knightly old Crusader, who sat his horse with lance in rest on a hill a little in the rear. "Come on!"

"I can't! I've caught a Tartar."

"Well, bring him along."

"He won't come."

"Well, then, come without him."

"He won't let me."

And so it often happened in the old days out here. When a man "caught" his bear and didn't have his gun he had to fight it out hand-to-hand. But fortunately, every man at all times had a knife in his belt. A knife never gets out of order, never "snaps," and a man in those days always had to have it with him to cut his food, cut brush, "crevice" for gold, and so on.

Oh! it is a grim picture to see a young fellow in his red shirt wheel about, when he can't run, thrust out his left hand, draw his knife with his right, and so, breast to breast, with the bear erect, strike and strike and strike to try to reach his heart before his left hand is eaten off to the elbow!

We have five kinds of bears in the Sierras. The "boxer," the "biter," the "hugger," are the most conspicuous. The other two are a sort of "all round" rough and tumble style of fighters.

The grizzly is the boxer. A game old beast he is, too, and would knock down all the John L. Sullivans you could put in the Sierras faster than you could set them up. He is a kingly old fellow and disdains familiarity. Whatever may be said to the contrary, he never "hugs" if he has room to box. In some desperate cases he has been known to bite, but ordinarily he obeys "the rules of the ring."

The cinnamon bear is a lazy brown brute, about one-half the size of the grizzly. He always insists on being very familiar, if not affectionate. This is the "hugger."

Next in order comes the big, sleek, black bear; easily tamed, too lazy to fight, unless forced to it. But when "cornered" he fights well, and, like a lion, bites to the bone.

After this comes the small and quarrelsome black bear with big ears, and a white spot on his breast. I have heard hunters say, but I don't quite believe it, that he sometimes points to this white spot on his breast as a sort of Free Mason's sign, as if to say, "Don't shoot." Next in order comes the smaller black bear with small ears. He is ubiquitous, as well as omniverous; gets into pig-pens, knocks over your beehives, breaks open your milk-house, eats more than two good-sized hogs ought to eat, and is off for the mountain top before you dream he is about. The first thing you see in the morning, however, will be some muddy tracks on the door steps. For he always comes and snuffles and shuffles and smells about the door in a good-natured sort of way, and leaves his card. The fifth member of the great bear family is not much bigger than an ordinary dog; but he is numerous, and he, too, is a nuisance.

Dog? Why not set the dog on him? Let me tell you. The California dog is a lazy, degenerate cur. He ought to be put with the extinct animals. He devotes his time and his talent to the flea. Not six months ago I saw a coon, on his way to my fish-pond in the pleasant moonlight, walk within two feet of my dog's nose and not disturb his slumbers.

We hope that it is impossible ever to have such a thing as hydrophobia in California. But as our dogs are too lazy to bite anything, we have thus far been unable to find out exactly as to that.

This last-named bear has a big head and small body; has a long, sharp nose and longer and sharper teeth than any of the others; he is a natural thief, has low instincts, carries his nose close to the ground, and, wherever possible, makes his road along on the mossy surface of fallen trees in humid forests. He eats fish—dead and decaying salmon—in such abundance that his flesh is not good in the salmon season.

It was with this last described specimen of the bear family that a precocious old boy who had hired out to some horse drovers, went in swimming years and years ago. The two drovers had camped to recruit and feed their horses on the wild grass and clover that grew at the headwaters of the Sacramento River, close up under the foot of Mount Shasta. A pleasant spot it was, in the pleasant summer weather.

This warm afternoon the two men sauntered leisurely away up Soda Creek to where their horses were grazing belly deep in grass and clover. They were slow to return, and the boy, as all boys will, began to grow restless. He had fished, he had hunted, had diverted himself in a dozen ways, but now he wanted something new. He got it.

A little distance below camp could be seen, through the thick foliage that hung and swung and bobbed above the swift waters, a long, mossy log that lay far out and far above the cool, swift river.

Why not go down through the trees and go out on that log, take off his clothes, dangle his feet, dance on the moss, do anything, everything that a boy wants to do?

In two minutes the boy was out on the big, long, mossy log, kicking his boots off, and in two minutes more he was dancing up and down on the humid, cool moss, and as naked as the first man, when he was first made.

And it was very pleasant. The great, strong river splashed and dashed and boomed below; above him the long green branches hung dense and luxuriant and almost within reach. Far off and away through their shifting shingle he caught glimpses of the bluest of all blue skies. And a little to the left he saw gleaming in the sun and almost overhead the everlasting snows of Mount Shasta.

Putting his boots and his clothes all carefully in a heap, that nothing might roll off into the water, he walked, or rather danced on out to where the further end of the great fallen tree lay lodged on a huge boulder in the middle of the swift and surging river. His legs dangled down and he patted his plump thighs with great satisfaction. Then he leaned over and saw some gold and silver trout, then he flopped over and lay down on his breast to get a better look at them. Then he thought he heard something behind him on the other end of the log! He pulled himself together quickly and stood erect, face about. There was a bear! It was one of those mean, sneaking, long-nosed, ant-eating little fellows, it is true, but it was a bear! And a bear is a bear to a boy, no matter about his size, age or character. The boy stood high up. The boy's bear stood up. And the boy's hair stood up!

The bear had evidently not seen the boy yet. But it had smelled his boots and clothes, and had got upon his dignity. But now, dropping down on all fours, with nose close to the mossy butt of the log, it slowly shuffled forward.

That boy was the stillest boy, all this time, that has ever been. Pretty soon the bear reached the clothes. He stopped, sat down, nosed them about as a hog might, and then slowly and lazily got up; but with a singular sort of economy of old clothes, for a bear, he did not push anything off into the river.

What next? Would he come any farther? Would he? Could he? Will he? The long, sharp little nose was once more to the moss and sliding slowly and surely toward the poor boy's naked shins. Then the boy shivered and settled down, down, down on his haunches, with his little hands clasped till he was all of a heap.

He tried to pray, but somehow or another, all he could think of as he sat there crouched down with all his clothes off was:

"Now I lay me down to sleep."

But all this could not last. The bear was almost on him in half a minute, although he did not lift his nose six inches till almost within reach of the boy's toes. Then the surprised bear suddenly stood up and began to look the boy in the face. As the terrified youth sprang up, he thrust out his left hand as a guard and struck the brute with all his might between the eyes with the other. But the left hand lodged in the two rows of sharp teeth and the boy and bear rolled into the river together.

But they were together only an instant. The bear, of course, could not breathe with his mouth open in the water, and so had to let go. Instinctively, or perhaps because his course lay in that direction, the bear struck out, swimming "dog fashion," for the farther shore. And as the boy certainly had no urgent business on that side of the river he did not follow, but kept very still, clinging to the moss on the big boulder till the bear had shaken the water from his coat and disappeared in the thicket.

Then the boy, pale and trembling from fright and the loss of blood, climbed up the broken end of the log, got his clothes, struggled into them as he ran, and so reached camp.

And he had not yelled! He tied up his hand in a piece of old flour sack, all by himself, for the men had not yet got back; and he didn't whimper! And what became of the boy? you ask.

The boy grew up as all energetic boys do; for there seems to be a sort of special providence for such boys.

And where is he now?

Out in California, trapping bear in the winter and planting olive trees in their season.

And do I know him?

Yes, pretty well, almost as well as any old fellow can know himself.

VI.

A FAT LITTLE EDITOR AND THREE LITTLE "BROWNS."

Mount Sinai, Heart of the Sierras—this place is one mile east and a little less than one mile perpendicular from the hot, dusty and dismal little railroad town down on the rocky banks of the foaming and tumbling Sacramento River. Some of the old miners are down there still—still working on the desolate old rocky bars with rockers. They have been there, some of them, for more than thirty years. A few of them have little orchards, or vineyards, on the steep, overhanging hills, but there is no home life, no white women to speak of, as yet. The battered and gray old miners are poor, lonely and discouraged, but they are honest, stout-hearted still, and of a much higher type than those that hang about the towns. It is hot down on the river—too hot, almost, to tell the truth. Even here under Mount Shasta, in her sheets of eternal snow, the mercury is at par.

This Mount Sinai is not a town; it is a great spring of cold water that leaps from the high, rocky front of a mountain which we have located as a summer home in the Sierras—myself and a few other scribes of California.

This is the great bear land. One of our party, a simple-hearted and honest city editor, who was admitted into our little mountain colony because of his boundless good nature and native goodness, had never seen a bear before he came here. City editors do not, as a rule, ever know much about bears. This little city editor is baldheaded, bow-legged, plain to a degree. And maybe that is why he is so good. "Give me fat men," said Caesar.

But give me plain men for good men, any time. Pretty women are to be preferred; but pretty men? Bah! I must get on with the bear, however, and make a long story a short story. We found our fat, bent-legged editor from the city fairly broiling in the little railroad town, away down at the bottom of the hill in the yellow golden fields of the Sacramento; and he was so limp and so lazy that we had to lay hold of him and get him out of the heat and up into the heart of the Sierras by main force.

Only one hour of climbing and we got up to where the little mountain streams come tumbling out of snow-banks on every side. The Sacramento, away down below and almost under us, from here looks dwindled to a brawling brook; a foamy white thread twisting about the boulders as big as meeting houses, plunging forward, white with fear, as if glad to get away—as if there was a bear back there where it came from. We did not register. No, indeed. This place here on Square Creek, among the clouds, where the water bursts in a torrent from the living rock, we have named Mount Sinai. We own the whole place for one mile square—the tall pine trees, the lovely pinewood houses; all, all. We proposed to hunt and fish, for food. But we had some bread, some bacon, lots of coffee and sugar. And so, whipping out our hooks and lines, we set off with the editor up a little mountain brook, and in less than an hour were far up among the fields of eternal snow, and finely loaded with trout.

What a bed of pine quills! What long and delicious cones for a camp fire! Some of those sugar-pine cones are as long as your arm. One of them alone will make a lofty pyramid of flame and illuminate the scene for half a mile about. I threw myself on my back and kicked up my heels. I kicked care square in the face. Oh, what freedom! How we would rest after dinner here! Of course we could not all rest or sleep at the same time. One of us would have to keep a pine cone burning all the time. Bears are not very numerous out here; but the California lion is both numerous and large here. The wild-cat, too, is no friend to the tourist. But we were not tourists. The land was and is ours. We would and all could defend our own.

The sun was going down. Glorious! The shades of night were coming up out of the gorges below and audaciously pursuing the dying sun. Not a sound. Not a sign of man or of beast. We were scattered all up and down the hill.

Crash! Something came tearing down the creek through the brush! The fat and simple-hearted editor, who had been dressing the homeopathic dose of trout, which inexperience had marked as his own, sprang up from the bank of the tumbling little stream above us and stood at his full height. His stout little knees for the first time smote together. I was a good way below him on the steep hillside. A brother editor was slicing bacon on a piece of reversed pine bark close by.

"Fall down," I cried, "fall flat down on your face."

It was a small she bear, and she was very thin and very hungry, with cubs at her heels, and she wanted that fat little city editor's fish. I know it would take volumes to convince you that I really meant for the bear to pass by him and come after me and my friend with both fish and bacon, and so, with half a line, I assert this truth and pass on. Nor was I in any peril in appropriating the little brown bear to myself. Any man who knows what he is about is as safe with a bear on a steep hillside as is the best bull-fighter in any arena. No bear can keep his footing on a steep hillside, much less fight. And whenever an Indian is in peril he always takes down hill till he comes to a steep plane, and then lets the bear almost overtake him, when he suddenly steps aside and either knifes the bear to the heart or lets the open-mouthed beast go on down the hill, heels over head.

The fat editor turned his face toward me, and it was pale. "What! Lie down and be eaten up while you lie there and kick up your heels and enjoy yourself? Never. We will die together!" he shouted.

He started for me as fast as his short legs would allow. The bear struck at him with her long, rattling claws. He landed far below me, and when he got up he hardly knew where he was or what he was. His clothes were in shreds, the back and bottom parts of them. The bear caught at his trout and was gone in an instant back with her two little cubs, and a moment later the little family had dined and was away, over the hill. She was a cinnamon bear, not much bigger than a big, yellow dog, and almost as lean and mean and hungry as any wolf could possibly be. We helped our inexperienced little friend slowly down to camp, forgetting all about the bacon and the fish till we came to the little board house, where we had coffee. Of course the editor could not go to the table now. He leaned, or rather sat, against a pine, drank copious cups of coffee and watched the stars, while I heaped up great piles of leaves and built a big fire, and so night rolled by in all her starry splendor as the men slept soundly all about beneath the lordly pines. But alas for the fat little editor; he did not like the scenery, and he would not stay. We saw him to the station on his way back to his little sanctum. He said he was satisfied. He had seen the "bar." His last words were, as he pulled himself close together in a modest corner in the car and smiled feebly: "Say, boys, you won't let it get in the papers, will you?"

VII.

TREEING A BEAR.

Away back in the "fifties" bears were as numerous on the banks of the Willamette River, in Oregon, as are hogs in the hickory woods of Kentucky in nut time, and that is saying that bears were mighty plenty in Oregon about forty years ago.

You see, after the missionaries established their great cattle ranches in Oregon and gathered the Indians from the wilderness and set them to work and fed them on beef and bread, the bears had it all their own way, till they literally overran the land. And this gave a great chance for sport to the sons of missionaries and the sons of new settlers "where rolls the Oregon."

And it was not perilous sport, either, for the grizzly was rarely encountered here. His home was further to the south. Neither was the large and clumsy cinnamon bear abundant on the banks of the beautiful Willamette in those dear old days, when you might ride from sun to sun, belly deep in wild flowers, and never see a house. But the small black bear, as indicated before, was on deck in great force, at all times and in nearly all places.

It was the custom in those days for boys to take this bear with the lasso, usually on horseback.

We would ride along close to the dense woods that grew by the river bank, and, getting between him and his base of retreat, would, as soon as we sighted a bear feeding out in the open plain, swing our lassos and charge him with whoop and yell. His habit of rearing up and standing erect and looking about to see what was the matter made him an easy prey to the lasso. And then the fun of taking him home through the long, strong grass!

As a rule, he did not show fight when once in the toils of the lasso; but in a few hours, making the best of the situation like a little philosopher, he would lead along like a dog.

There were, of course, exceptions to this exemplary conduct.

On one occasion particularly, Ed Parish, the son of a celebrated missionary, came near losing his life by counting too confidently on the docility of a bear which he had taken with a lasso and was leading home.

His bear suddenly stopped, stood up and began to haul in the rope, hand over hand, just like a sailor. And as the other end of the rope was fastened tightly to the big Spanish pommel of the saddle, why of course the distance between the bear and the horse soon grew perilously short, and Ed Parish slid from his horse's back and took to the brush, leaving horse and bear to fight it out as best they could.

When he came back, with some boys to help him, the horse was dead and the bear was gone, having cut the rope with his teeth.

After having lost his horse in this way, poor little Ed Parish had to do his hunting on foot, and, as my people were immigrants and very poor, why we, that is my brother and I, were on foot also. This kept us three boys together a great deal, and many a peculiar adventure we had in those dear days "when all the world was young."

Ed Parish was nearly always the hero of our achievements, for he was a bold, enterprising fellow, who feared nothing at all. In fact, he finally lost his life from his very great love of adventure. But this is too sad to tell now, and we must be content with the story about how he treed a bear for the present.

We three boys had gone bear hunting up a wooded canyon near his father's ranch late one warm summer afternoon. Ed had a gun, but, as I said before, my people were very poor, so neither brother nor I as yet had any other arms or implements than the inseparable lasso.

Ed, who was always the captain in such cases, chose the center of the dense, deep canyon for himself, and, putting my brother on the hillside to his right and myself on the hillside to his left, ordered a simultaneous "Forward march."

After a time we heard him shoot. Then we heard him shout. Then there was a long silence.

Then suddenly, high and wild, his voice rang out through the tree tops down in the deep canyon.

"Come down! Come quick! I've treed a bear! Come and help me catch him; come quick! Oh, Moses! come quick, and—and—and catch him!"

My brother came tearing down the steep hill on his side of the canyon as I descended from my side. We got down about the same time, but the trees in their dense foliage, together with the compact underbrush, concealed everything. We could see neither bear nor boy.

This Oregon is a damp country, warm and wet; nearly always moist and humid, and so the trees are covered with moss. Long, gray, sweeping moss swings from the broad, drooping boughs of fir and pine and cedar and nearly every bit of sunlight is shut out in these canyons from one year's end to the other. And it rains here nearly half of the year; and then these densely wooded canyons are as dark as caverns. I know of nothing so grandly gloomy as these dense Oregon woods in this long rainy season.

I laid my ear to the ground after I got a glimpse of my brother on the other side of the canyon, but could hear nothing at all but the beating of my heart.

Suddenly there was a wild yell away up in the dense boughs of a big mossy maple tree that leaned over toward my side of the canyon. I looked and looked with eagerness, but could see nothing whatever.

Then again came the yell from the top of the big leaning maple. Then there was a moment of silence, and then the cry: "Oh, Moses! Why don't you come, I say, and help me catch him?" By this time I could see the leaves rustling. And I could see the boy rustling, too.

And just behind him was a bear. He had treed the bear, sure enough!

My eyes gradually grew accustomed to the gloom and density, and I now saw the red mouth of the bear amid the green foliage high overhead. The bear had already pulled off one of Ed's boots and was about making a bootjack of his big red mouth for the other.

"Why don't you come on, I say, and help me catch him?"

He kicked at the bear, and at the same time hitched himself a little further along up the leaning trunk, and in doing so kicked his remaining boot into the bear's mouth.

"Oh, Moses, Moses! Why don't you come? I've got a bear, I tell you."

"Where is it, Ed?" shouted my brother on the other side.

But Ed did not tell him, for he had not yet got his foot from the bear's mouth, and was now too busy to do anything else but yell and cry "Oh, Moses!"

Then my brother and I shouted out to Ed at the same time. This gave him great courage. He said something like "Confound you!" to the bear, and getting his foot loose without losing the boot he kicked the bear right on the nose. This brought things to a standstill. Ed hitched along a little higher up, and as the leaning trunk of the tree was already bending under his own and the bear's weight, the infuriated brute did not seem disposed to go further. Besides, as he had been mortally wounded, he was probably growing too weak to do much now.

My brother got to the bottom of the canyon and brought Ed's gun to where I stood. But, as we had no powder or bullets, and as Ed could not get them to us, even if he would have been willing to risk our shooting at the bear, it was hard to decide what to do. It was already dusk and we could not stay there all night.

"Boys," shouted Ed, at last, as he steadied himself in the forks of a leaning and overhanging bough, "I'm going to come down on my laz rope. There, take that end of it, tie your laz ropes to it and scramble up the hill."

We obeyed him to the letter, and as we did so, he fastened his lasso firmly to the leaning bough and descended like a spider to where we had stood a moment before. We all scrambled up out of the canyon together and as quickly as possible.

When we went back next day to get our ropes we found the bear dead near the root of the old mossy maple. The skin was a splendid one, and Ed insisted that my brother and I should have it, and we gladly accepted it.

My brother, who was older and wiser than I, said that he made us take the skin so that we would not be disposed to tell how he had "treed a bear." But I trust not, for he was a very generous-hearted fellow. Anyhow, we never told the story while he lived.

VIII.

BILL CROSS AND HIS PET BEAR.

When my father settled down at the foot of the Oregon Sierras with his little family, long, long years ago, it was about forty miles from our place to the nearest civilized settlement.

People were very scarce in those days, and bears, as said before, were very plenty. We also had wolves, wild-cats, wild cattle, wild hogs, and a good many long-tailed and big-headed yellow Californian lions.

The wild cattle, brought there from Spanish Mexico, next to the bear, were most to be feared. They had long, sharp horns and keen, sharp hoofs. Nature had gradually helped them out in these weapons of defense. They had grown to be slim and trim in body, and were as supple and swift as deer. They were the deadly enemies of all wild beasts; because all wild beasts devoured their young.

When fat and saucy, in warm summer weather, these cattle would hover along the foothills in bands, hiding in the hollows, and would begin to bellow whenever they saw a bear or a wolf, or even a man or boy, if on foot, crossing the wide valley of grass and blue camas blossoms. Then there would be music! They would start up, with heads and tails in the air, and, broadening out, left and right, they would draw a long bent line, completely shutting off their victim from all approach to the foothills. If the unfortunate victim were a man or boy on foot, he generally made escape up one of the small ash trees that dotted the valley in groves here and there, and the cattle would then soon give up the chase. But if it were a wolf or any other wild beast that could not get up a tree, the case was different. Far away, on the other side of the valley, where dense woods lined the banks of the winding Willamette river, the wild, bellowing herd would be answered. Out from the edge of the woods would stream, right and left, two long, corresponding, surging lines, bellowing and plunging forward now and then, their heads to the ground, their tails always in the air and their eyes aflame, as if they would set fire to the long gray grass. With the precision and discipline of a well-ordered army, they would close in upon the wild beast, too terrified now to either fight or fly, and, leaping upon him, one after another, with their long, sharp hoofs, he would, in a little time, be crushed into an unrecognizable mass. Not a bone would be left unbroken. It is a mistake to suppose that they ever used their long, sharp horns in attack. These were used only in defense, the same as elk or deer, falling on the knees and receiving the enemy on their horns, much as the Old Guard received the French in the last terrible struggle at Waterloo.

Bill Cross was a "tender foot" at the time of which I write, and a sailor, at that. Now, the old pilgrims who had dared the plains in those days of '49, when cowards did not venture and the weak died on the way, had not the greatest respect for the courage or endurance of those who had reached Oregon by ship. But here was this man, a sailor by trade, settling down in the interior of Oregon, and, strangely enough, pretending to know more about everything in general and bears in particular than either my father or any of his boys!

He had taken up a piece of land down in the pretty Camas Valley where the grass grew long and strong and waved in the wind, mobile and beautiful as the mobile sea.

The good-natured and self-complacent old sailor liked to watch the waving grass. It reminded him of the sea, I reckon. He would sometimes sit on our little porch as the sun went down and tell us boys strange, wild sea stories. He had traveled far and seen much, as much as any man can see on water, and maybe was not a very big liar, for a sailor, after all. We liked his tales. He would not work, and so he paid his way with stories of the sea. The only thing about him that we did not like, outside of his chronic idleness, was his exalted opinion of himself and his unconcealed contempt for everybody's opinion but his own.

"Bill," said my father one day, "those black Spanish cattle will get after that red sash and sailor jacket of yours some day when you go down in the valley to your claim, and they won't leave a grease spot. Better go horseback, or at least take a gun, when you go down next time."

"Pshaw! Squire. I wish I had as many dollars as I ain't afeard of all the black Spanish cattle in Oregon. Why, if they're so blasted dangerous, how did your missionaries ever manage to drive them up here from Mexico, anyhow?"

Still, for all that, the very next time that he saw the old sailor setting out at his snail pace for his ranch below, slow and indolent as if on the deck of a ship, my father insisted that he should go on horseback, or at least take a gun.

"Pooh, pooh! I wouldn't be bothered with a horse or a gun. Say, I'm goin' to bring your boys a pet bear some day."

And so, cocking his little hat down over his right eye and thrusting his big hands into his deep pockets almost to the elbows, he slowly and lazily whistled himself down the gradual slope of the foothills, waist deep in the waving grass and delicious wild flowers, and soon was lost to sight in the great waving sea.

Two things may be here written down. He wouldn't ride a horse because he couldn't, and for the same reason he wouldn't use a gun. Again let it be written down, also, that the reason he was going away that warm autumn afternoon was that there was some work to do. These facts were clear to my kind and indulgent father; but of course we boys never thought of it, and laid our little shoulders to the hard work of helping father lift up the long, heavy poles that were to complete the corral around our pioneer log cabin, and we really hoped and half believed that he might bring home a little pet bear.

This stout log corral had become an absolute necessity. It was high and strong, and made of poles or small logs stood on end in a trench, after the fashion of a primitive fort or stout stockade. There was but one opening, and that was a very narrow one in front of the cabin door. Here it was proposed to put up a gate. We also had talked about port-holes in the corners of the corral, but neither gate nor port-holes were yet made. In fact, as said before, the serene and indolent man of the sea always slowly walked away down through the grass toward his untracked claim whenever there was anything said about port-holes, posts or gates.

Father and we three little boys had only got the last post set and solidly "tamped" in the ground as the sun was going down.

Suddenly we heard a yell; then a yelling, then a bellowing. The yelling was heard in the high grass in the Camas Valley below, and the bellowing of cattle came from the woody river banks far beyond.

Then up on the brown hills of the Oregon Sierras above us came the wild answer of the wild black cattle of the hills, and a moment later, right and left, the long black lines began to widen out; then down they came, like a whirlwind, toward the black and surging line in the grass below. We were now almost in the center of what would, in a little time, be a complete circle and cyclone of furious Spanish cattle.

And now, here is something curious to relate. Our own cows, poor, weary, immigrant cows of only a year before, tossed their tails in the air, pawed the ground, bellowed and fairly went wild in the splendid excitement and tumult. One touch of nature made the whole cow world kin!

Father clambered up on a "buck-horse" and looked out over the stockade; and then he shouted and shook his hat and laughed as I had never heard him laugh before. For there, breathless, coatless, hatless, came William Cross, Esq., two small wolves and a very small black bear! They were all making good time, anywhere, anyway, to escape the frantic cattle. Father used to say afterwards, when telling about this little incident, that "it was nip and tuck between the four, and hard to say which was ahead." The cattle had made quite a "round-up."

They all four straggled in at the narrow little gate at about the same time, the great big, lazy sailor in a hurry, for the first time in his life.

But think of the coolness of the man, as he turned to us children with his first gasp of breath, and said, "Bo—bo—boys, I've bro—bro—brought you a little bear!"

The wolves were the little chicken thieves known as coyotes, quite harmless, as a rule, so far as man is concerned, but the cattle hated them and they were terrified nearly to death.

The cattle stopped a few rods from the stockade. We let the coyotes go, but we kept the little bear and named him Bill Cross. Yet he was never a bit cross, despite his name.

IX.

THE GREAT GRIZZLY BEAR.

(Ursus Ferox.)

"The Indians have unbounded reverence for this bear. When they kill one, they make exculpating speeches to it, smoke tobacco to it, call it grandfather, ancestor, etc."

P. Martin Duncan, M. B., F. R. S., F. G. S.

Kings College, London.

The Indians with whom I once lived in the Californian Sierras held the grizzly bear in great respect and veneration. Some writers have said that this was because they were afraid of this terrible king of beasts. But this is not true. The Indian, notwithstanding his almost useless bow and arrow in battles with this monster, was not controlled by fear. He venerated the grizzly bear as his paternal ancestor. And here I briefly set down the Modoc and Mount Shasta Indians' account of their own creation.

They, as in the Biblical account of the creation of all things, claim to have found the woods, wild beasts, birds and all things waiting for them, as did Adam and Eve.

The Indians say the Great Spirit made this mountain first of all. Can you not see how it is? they say. He first pushed down snow and ice from the skies through a hole which he made in the blue heavens by turning a stone round and round, till he made this great mountain; then he stepped out of the clouds onto the mountain-top, and descended and planted the trees all around by putting his finger on the ground. The sun melted the snow, and the water ran down and nurtured the trees and made the rivers. After that he made the fish for the rivers out of the small end of his staff. He made the birds by blowing some leaves, which he took up from the ground, among the trees. After that he made the beasts out of the remainder of his stick, but made the grizzly bear out of the big end, and made him master over all the others. He made the grizzly so strong that he feared him himself, and would have to go up on top of the mountain out of sight of the forest to sleep at night, lest the grizzly, who, as will be seen, was much more strong and cunning then than now, should assail him in his sleep. Afterwards, the Great Spirit, wishing to remain on earth and make the sea and some more land, converted Mount Shasta, by a great deal of labor, into a wigwam, and built a fire in the center of it and made it a pleasant home. After that, his family came down, and they all have lived in the mountain ever since. They say that before the white man came they could see the fire ascending from the mountain by night and the smoke by day, every time they chose to look in that direction. They say that one late and severe springtime, many thousand snows ago, there was a great storm about the summit of Mount Shasta, and that the Great Spirit sent his youngest and fairest daughter, of whom he was very fond, up to the hole in the top, bidding her to speak to the storm that came up from the sea, and tell it to be more gentle or it would blow the mountain over. He bade her do this hastily, and not put her head out, lest the wind should catch her in the hair and blow her away. He told her she should only thrust out her long red arm and make a sign, and then speak to the storm without.

The child hastened to the top and did as she was bid, and was about to return, but having never yet seen the ocean, where the wind was born and made his home, when it was white with the storm, she stopped, turned and put her head out to look that way, when lo! the storm caught in her long red hair, and blew her out and away down and down the mountain side. Here she could not fix her feet in the hard, smooth ice and snow, and so slid on and on down to the dark belt of firs below the snow rim.

Now, the grizzly bears possessed all the wood and all the land down to the sea at that time, and were very numerous and very powerful. They were not exactly beasts then, although they were covered with hair, lived in caves and had sharp claws; but they walked on two feet, and talked, and used clubs to fight with, instead of their teeth and claws, as they do now.

At this time, there was a family of grizzlies living close up to the snows. The mother had lately brought forth, and the father was out in quest of food for the young, when, as he returned with his club on his shoulder and a young elk in his left hand, under his arm, he saw this little child, red like fire, hid under a fir-bush, with her long hair trailing in the snows, and shivering with fright and cold. Not knowing what to make of her, he took her to the old mother, who was very learned in all things, and asked her what this fair and frail thing was that he had found shivering under a fir-bush in the snow. The old mother grizzly, who had things pretty much her own way, bade him leave the child with her, but never mention it to anyone, and she would share her breast with her, and bring her up with the other children, and maybe some great good would come of it.

The old mother reared her as she promised to do, and the old hairy father went out every day, with his club on his shoulder, to get food for his family, till they were all grown up and able to do for themselves.

"Now," said the old mother Grizzly to the old father Grizzly, as he stood his club by the door and sat down one day, "our oldest son is quite grown up and must have a wife. Now, who shall it be but the little red creature you found in the snow under the black fir-bush." So the old father Grizzly kissed her, said she was very wise, then took up his club on his shoulder and went out and killed some meat for the marriage feast.

They married and were very happy, and many children were born to them. But, being part of the Great Spirit and part of the grizzly bear, these children did not exactly resemble either of their parents, but partook somewhat of the nature and likeness of both. Thus was the red man created; for these children were the first Indians.

All the other grizzlies throughout the black forests, even down to the sea, were very proud and very kind, and met together, and, with their united strength, built for the lovely little red princess a wigwam close to that of her father, the Great Spirit. This is what is now called "Little Mount Shasta."

After many years, the old mother Grizzly felt that she soon must die, and, fearing that she had done wrong in detaining the child of the Great Spirit, she could not rest till she had seen him and restored to him his long-lost treasure and asked his forgiveness.

With this object in view, she gathered together all the grizzlies at the new and magnificent lodge built for the princess and her children, and then sent her eldest grandson to the summit of Mount Shasta in a cloud, to speak to the Great Spirit and tell him where he could find his long-lost daughter.

When the Great Spirit heard this, he was so glad that he ran down the mountain side on the south so fast and strong that the snow was melted off in places, and the tokens of his steps remain to this day. The grizzlies went out to meet him by thousands; and as he approached they stood apart in two great lines, with their clubs under their arms, and so opened a lane through which he passed in great state to the lodge where his daughter sat with her children.

But when he saw the children, and learned how the grizzlies that he had created had betrayed him into the creation of a new race, he was very wroth, and frowned on the old mother Grizzly till she died on the spot. At this, the grizzlies all set up a dreadful howl; but he took his daughter on his shoulder and, turning to all the grizzlies, bade them hold their tongues, get down on their hands and knees and so remain till he returned. They did as they were bid, and he closed the door of the lodge after him, drove all the children out into the world, passed out and up the mountain and never returned to the timber any more.

So the grizzlies could not rise up any more, or make a noise, or use their clubs, but ever since have had to go on all-fours, much like other beasts, except when they have to fight for their lives; then the Great Spirit permits them to stand up and fight with their fists like men.

That is why the Indians about Mount Shasta will never kill or interfere in any way with a grizzly. Whenever one of their number is killed by one of these kings of the forest, he is burned on the spot, and all who pass that way for years cast a stone on the place till a great pile is thrown up. Fortunately, however, grizzlies are not now plentiful about the mountain.

In proof of the story that the grizzly once stood and walked erect and was much like a man, they show that he has scarcely any tail, and that his arms are a great deal shorter than his legs, and that they are more like a man than any other animal.

X.

AS A HUMORIST.

Not long ago, about the time a party of Americans were setting out for India to hunt the tiger, a young banker from New York came to California to hunt what he rightly considered the nobler beast.

He chartered a small steamer in San Francisco Bay and taking with him a party of friends, as well as a great-grandson of Daniel Boone, a famous hunter, for a guide, he sailed up the coast to the redwood wilderness of Humboldt. Here he camped on the bank of a small stream in a madrona thicket and began to hunt for his bear. He found his bear, an old female with young cubs. As Boone was naturally in advance when the beast was suddenly stumbled upon, he had to do the fighting, and this gave the banker from the States a chance to scramble up a small madrona. Of course he dropped his gun. They always do drop their guns, by some singularly sad combination of accidents, when they start up a tree with two rows of big teeth in the rear, and it is hardly fair to expect the young bear-hunter from New York to prove an exception. Poor Boone was severely maltreated by the savage old mother grizzly in defense of her young. There was a crashing of brush and a crushing of bones, and then all was still.

Of course he dropped his gun.

Suddenly the bear seemed to remember that there was a second party who had been in earnest search for a bear, and looking back down the trail and up in the boughs of a small tree, she saw a pair of boots. She left poor Boone senseless on the ground and went for those boots. Coming forward, she reared up under the tree and began to claw for the capitalist. He told me that she seemed to him, as she stood there, to be about fifty feet high. Then she laid hold of the tree.

Fortunately this madrona tree is of a hard and unyielding nature, and with all her strength she could neither break nor bend it. But she kept thrusting up her long nose and longer claws, laying hold first of his boots, which she pulled off, one after the other, with her teeth, then with her claws she took hold of one garment and then another till the man of money had hardly a shred, and his legs were streaming with blood. Fearing that he should faint from loss of blood, he lashed himself to the small trunk of the tree by his belt and then began to scream with all his might for his friends.

When the bear became weary of clawing up at the dangling legs she went back and began to turn poor Boone over to see if he showed any signs of life. Then she came back and again clawed a while at the screaming man up the madrona tree. It was great fun for the bear!

To cut a thrilling story short, the party in camp on the other side of the creek finally came in hail, when the old bear gathered up her babies and made safe exit up a gulch. Boone, now in Arizona, was so badly crushed and bitten that his life was long despaired of, but he finally got well. The bear, he informed me, showed no disposition to eat him while turning him over and tapping him with her foot and thrusting her nose into his bleeding face to see if he still breathed.

Story after story of this character could be told to prove that the grizzly at home is not entirely brutal and savage; but rather a good-natured lover of his family and fond of his sly joke.

XI.

A GRIZZLY'S SLY LITTLE JOKE.

I know an old Indian who was terribly frightened by an old monster grizzly and her half-grown cub, one autumn, while out gathering manzanita berries. But badly as he was frightened, he was not even scratched.

It seems that while he had his head raised, and was busy gathering and eating berries, he almost stumbled over an old bear and her cub. They had eaten their fill and fallen asleep in the trail on the wooded hillside. The old Indian had only time to turn on his heel and throw himself headlong in the large end of a hollow log, which luckily lay at hand. This, however, was only a temporary refuge. He saw, to his delight, that the log was open at the other end, and corkscrewing his way along toward the further end, he was about to emerge, when, to his dismay, he saw the old mother sitting down quietly waiting for him!

After recovering his breath as best he could in his hot and contracted quarters, he elbowed and corkscrewed himself back to the place by which he first entered. But lo! the bear was there, sitting down, half smiling, and waiting to receive him warmly. This, the old Indian said, was repeated time after time, till he had no longer strength left to struggle further, and turned on his face to die, when she put her head in, touched the top of his head gently with her nose and then drew back, took her cub with her and shuffled on.

I went to the spot with the Indian a day or two afterward, and am convinced that his story was exactly as narrated. And when you understand that the bear could easily have entered the hollow log and killed him at any time, you will see that she had at least a faint sense of fun in that "cat and mouse" amusement with the frightened Indian.

The bear was waiting there.

XII.

THE GRIZZLY AS FREMONT FOUND HIM.

General Fremont found this powerful brute to be a gregarious and confiding creature, fond of his family and not given to disturbing those who did not disturb him. In his report to the government—1847—he tells of finding a large family of grizzly bears gathering acorns very much as the native Indians gathered them, and this not far from a small Mexican town. He says that riding at the head of his troops he saw, on reaching the brow of a little grassy hill set with oaks, a great commotion in the boughs of one of the largest trees, and, halting to cautiously reconnoiter, he noticed that there were grouped about the base of the tree and under its wide boughs, several huge grizzlies, employed in gathering and eating the acorns which the baby grizzlies threw down from the thick branches overhead. More than this, he reports that the baby bears, on seeing him, became frightened, and attempted to descend to the ground and run away, but the older bears, which had not yet discovered the explorers, beat the young ones and drove them back up the tree, and compelled them to go on with their work, as if they had been children.

In the early '50s, I, myself, saw the grizzlies feeding together in numbers under the trees, far up the Sacramento Valley, as tranquilly as a flock of sheep. A serene, dignified and very decent old beast was the full-grown grizzly as Fremont and others found him here at home. This king of the continent, who is quietly abdicating his throne, has never been understood. The grizzly was not only every inch a king, but he had, in his undisputed dominion, a pretty fair sense of justice. He was never a roaring lion. He was never a man-eater. He is indebted for his character for ferocity almost entirely to tradition, but, in some degree, to the female bear when seeking to protect her young. Of course, the grizzlies are good fighters, when forced to it; but as for lying in wait for anyone, like the lion, or creeping, cat-like, as the tiger does, into camp to carry off someone for supper, such a thing was never heard of in connection with the grizzly.

The grizzly went out as the American rifle came in. I do not think he retreated. He was a lover of home and family, and so fell where he was born. For he is still found here and there, all up and down the land, as the Indian is still found, but he is no longer the majestic and serene king of the world. His whole life has been disturbed, broken up; and his temper ruined. He is a cattle thief now, and even a sheep thief. In old age, he keeps close to his canyon by day, deep in the impenetrable chaparral, and at night shuffles down hill to some hog-pen, perfectly careless of dogs or shots, and, tearing out a whole side of the pen, feeds his fill on the inmates.

One of the interior counties kept a standing reward for the capture of an old grizzly of this character for several years. But he defied everything and he escaped everything but old age. Some hunters finally crept in to where the old king lay, nearly blind and dying of old age, and dispatched him with a volley from several Winchester rifles. It was found that he was almost toothless, his paws had been terribly mutilated by numerous steel traps, and it is said that his kingly old carcass had received nearly lead enough to sink a small ship. There were no means of ascertaining his exact weight, but it was claimed that skin, bone and bullets, as he was found, he would have weighed well nigh a ton.

XIII.

THE BEAR WITH SPECTACLES.

And now let us go down to near the mouth of the Father of Waters, to "Barra Tarra Land" or Barren Land, as it was called of old by Cervantes, in the kingdom of Sancho Panza. Strange how little the great men of the old world knew of this new world! In one of his plays Shakespeare speaks of ships from Mexico; in another he means to mention the Bermudas. Burns speaks of a Newfoundland dog as

> "Whelped in a country far abroad
>
> Where boatmen gang to fish for cod,"

and Byron gets in a whole lot about Daniel Boone; but as a rule we were ignored.

Barra Tarra, so called, is the very richest part of this globe. It must have been rich always, rich as the delta of the Nile; but now, with the fertility of more than a dozen States dumped along there annually, it is rich as cream is rich.

The fish, fowl, oysters of Barra Tarra—ah, the oysters! No oysters in the world like these for flavor, size and sweetness. They are so enormous in size that—but let me illustrate their size by an anecdote of the war.

A Yankee captain, hungry and worn out hewing his way with his sword from Chicago to the sea, as General Logan had put it, sat down in a French restaurant in New Orleans, and while waiting for a plate of the famous Barra Tarra raw oysters, saw that a French creole sitting at the same little side table was turning over and over with his fork a solitary and most tempting oyster of enormous size, eyeing it ruefully.

"Why don't you eat him?"

"By gar! I find him too big for me. You like?"

"Certainly. Not too big for me. See this!" and snatching the fork from the Frenchman the oyster was gone at a gulp.

The little Frenchman shrugged his shoulders, looked at the gallant officer a moment and then said in a fit of enthusiastic admiration:

"By gar, Monsieur Capitaine, you are one mighty brave man! I did try him t'ree times zat way, but he no stay."

The captain threw up his arms and—his oyster!—so runs the story.

The soil along the river bank is so rich that weeds, woods, vines, trench close and hard on the heels of the plowman. A plantation will almost perish from the earth, as it were, by a few years of abandonment. And so it is that you see miles and miles on either side—parishes on top of parishes, in fact—fast returning to barbarism, dragging the blacks by thousands down to below the level of brutes with them, as you descend from New Orleans toward the mouth of the mighty river, nearly one hundred miles from the beautiful "Crescent City." And, ah, the superstition of these poor blacks!

You see hundreds of little white houses, old "quarters," and all tenantless now, save one or two on each plantation. Cheap sugar and high wages, as compared with old times of slavery—but then the enormous cost of keeping up the levees, and above all, the continued peril to life and property, with a mile of swift, muddy water sweeping seaward high above your head—these things are making a desert of the richest lands on earth. We are gaining ground in the West, but we are losing ground in the South, the great, silent South.

Of course, the world, we, civilization, will turn back to this wondrous region some day, when we have settled the West; for the mouth of the mightiest river on the globe is a fact; it is the mouth by which this young nation was trained in its younger days, and we cannot ignore it in the end, however willing we may be to do so now.

Strange how wild beasts and all sorts of queer creatures are overrunning the region down there, too, growing like weeds, increasing as man decreases. I found a sort of marsh bear here. He looks like the sloth bear (Ursus Labiatus) of the Ganges, India, as you see him in the Zoo of London, only he is not a sloth, by any means. The negroes are superstitiously afraid of him, and their dogs, very numerous, and good coon dogs, too, will not touch him. His feet are large and flat, to accommodate him in getting over the soft ground, while his shaggy and misshapen body is very thin and light. His color is as unlovely as his shape—a sort of faded, dirty brown or pale blue, with a rim of dirty white about the eyes that makes him look as if he wore spectacles when he stops and looks at you.

As he is not fit to eat because he lives on fish and oysters, sportsmen will not fire at him; and as the poor, superstitious, voodoo-worshiping negroes, and their dogs, too, run away as soon as he is seen, he has quite a habit of stopping and looking at you through his queer spectacles as long as you are in sight. He looks to be a sort of second-hand bear, his shaggy, faded, dirty coat of hair looking as if he had been stuffed, like an old sofa, with the stuffing coming out—a very second-hand appearance, to be sure.

Now, as I have always had a fondness for skins—having slept on them and under them all my life, making both bed and carpet of them—I very much wanted a skin of this queer marsh bear which the poor negroes both adore and dread as a sort of devil. But, as no one liked him well enough to kill him, I must do it myself; and with this object, along with my duty to describe the drowning plantations, I left New Orleans with Colonel Bloom, two good guns, and something to eat and to drink, and swept down the great river to the landing in the outer edge of the timber belt.

And how strange this landing! As a rule you have to climb up to the shore from a ship. Here, after setting foot on the levee, we walked down, down, down to reach the level land—a vast field of fevers.

I had a letter of introduction to the "preacher." He was a marvel of rags, preached every day and night, up and down the river, and received 25 cents a day from the few impoverished white planters, too poor to get away, for his influence for good among the voodoo blacks. Not that they could afford to care for the negroes, those few discouraged and fever-stricken planters on their plantations of weeds and water, but they must, now and then, have these indolent and retrograding blacks to plant or cut down their cane, or sow and gather their drowning patches of rice, and the preacher could preach them into working a little, when right hungry.

The ragged black took my letter and pretended to read it. Poor fellow, he could not read, but pride, or rather vanity, made him act a lie. Seeing the fact, I contrived to tell him that it was from a colored clergyman, and that I had come to get him and his dogs to help me kill a bear. The blacks now turned white; or at least white around the lips. The preacher shuddered and shrugged his shoulders and finally groaned in his grief.

Let us omit the mosquitoes, the miserable babies, nude as nature, and surely very hungry in this beauteous place of fertility. They hung about my door, a "quarters" cabin with grass knee high through the cracks in the floor, like flies, till they got all my little store of supplies, save a big flask of "provisions" which General Beauregard had given me for Colonel Bloom, as a preventive against the deadly fever. No, it was not whiskey, not all whiskey, at least, for it was bitter as gall with quinine. I had to help the Colonel sample it at first, but I only helped him sample it once. It tasted so vilely that it seemed to me I should, as between the two, prefer fever.

And such a moon! The ragged minister
 stood whooping up his numerous dogs and gathering his sullen clan of blacks to get that bear and that promised $5.

Away from up toward New Orleans, winding, sweeping, surging, flashing like a mighty sword of silver, the Father of Waters came through the air, high above our heads and level with the topmost limit of his artificial banks. The blacks were silent, ugly, sullen, and so the preacher asked for and received the five silver dollars in advance. This made me suspicious, and, out of humor, I went into my cabin and took Colonel Bloom into a corner and told him what had been done. He did not say one word but took a long drink of preventive against the fever, as General Beauregard had advised and provided.

Then we set out for the woods, through weeds that reached to our shoulders, the negroes in a string, slow, silent, sullen and ugly, the brave bear dogs only a little behind the negroes. The preacher kept muttering a monotonous prayer.

But that moon and that mighty sword of silver in the air, the silence, the large solemnity, the queer line of black heads barely visible above the sea of weeds! I was not right certain that I had lost any bear as we came to the edge of the moss-swept cypress woods, for here the negroes all suddenly huddled up and muttered and prayed with one voice. Aye, how they prayed in their piteous monotone! How sad it all was!

The dogs had sat down a few rods back, a line of black dots along the path through the tall weeds, and did not seem to care for anything at all. I had to lay my hand on the preacher's shoulder and ask him to please get on; then they all started on together, and oh, the moon, through the swaying cypress moss, the mighty river above!

It was with great effort that I got them to cross a foot-log that lay across a lagoon only a little way in the moss-hung woods, the brave dogs all the time only a short distance behind us still. It was a hot night and the mosquitoes were terrible in the woods, but I doubt if they bite the blacks as they did me. Surely not, else they would not be even as nearly alive as they are.

Having got them across the lagoon, I gave them each 25 cents more, and this made them want to go home. The dogs had all sat down in a queer row on the foot-log. Such languor, such laziness, such idiotic helplessness I never saw before, even on the Nile. The blacks, as well as the dogs, seemed to be afraid to move now. The preacher again began to mumble a prayer, and the whole pack with him; and then they prayed again, this time not so loudly. And although there was melody of a sort in their united voices, I am certain they used no words, at least no words of any real language.

Suddenly the dogs got up and came across and hid among the men, and the men huddled up close; for right there on the other end of the log, with his broad right foot resting on it, was the shaggy little beast we were hunting for. We had found our bear, or rather, he had found us, and it was clear that he meant to come over and interview us at once.

The preacher crouched behind me as I cocked and raised my gun, the blacks hid behind the preacher, and I think, though I had not time to see certainly, that the dogs hid behind the blacks.

I fired at the dim white spot on the bear's breast and sent shot after shot into his tattered coat, for he was not ten lengths of an old Kentucky ramrod distant, and he fell dead where he stood, and I went over and dragged him safely up on the higher bank.

Then the wild blacks danced and sang and sang and danced, till one of them slipped and fell into the lagoon. They fished him out and all returned to where I was, with the dead bear, dogs and all in great good spirits. Tying the bear's feet together with a withe they strung him on a pole and we all went back home, the blacks singing all the way some barbaric half French song at the top of their melodious voices.

But Colonel Bloom was afraid that the one who had fallen in the river might take the fever, and so as soon as we got safe back he drank what was left in the bottle General Beauregard had sent him and he went to sleep; while the superstitious blacks huddled together under the great levee and skinned the bear in the silver moonlight, below the mighty river. I gave them each a silver dollar—very bright was the brand new silver from the mint of New Orleans, but not nearly so bright as the moon away down there by the glowing rim of the Mexican seas where the spectacled bear abides in the classic land, Barra Tarra, Kingdom of Sancho Panza.

XIV.

THE BEAR-SLAYER OF SAN DIEGO.

Let us now leave the great grizzly and the little marsh bear in spectacles behind us and tell about a boy, a bear-slayer; not about a bear, mind you. For the little fish-eating black bear which he killed and by which he got his name is hardly worth telling about. This bear lives in the brush along the sea-bank on the Mexican and Southern California coast and has huge feet but almost no hair. I don't know any name for him, but think he resembles the "sun bear" (Ursus Titanus) more than any other. His habit of rolling himself up in a ball and rolling down hill after you is like that of the porcus or pig bear.

You may not know that a bear, any kind of a bear, finds it hard work running down hill, because of his short arms, so when a man who knows anything about bears is pursued, or thinks he is pursued, he always tries, if he knows himself, to run down hill. A man can escape almost any bear by running down hill, except this little fellow along the foothills by the Mexican seas. You see, he has good bear sense, like the rest of the bear family, and gets along without regard to legs of any sort, sometimes.

This boy that I am going to tell about was going to school on the Mexican side of the line between the two republics, near San Diego, California, when a she bear which had lost her cub caught sight of the boys at play down at the bottom of a high, steep hill, and she rolled for them, rolled right among the little, half-naked fellows, and knocked numbers of them down. But before she could get the dust out of her eyes and get up, this boy jumped on her and killed her with his knife.

The governor remembered the boy for his pluck and presence of mind and he was quite a hero and was always called "The Bear-Slayer" after that.

Some rich ladies from Boston, hearing about his brave act, put their heads together and then put their hands in their pockets and sent him to a higher school, where the following incident took place.

I ought to mention that this little Mexican bear, though he has but little hair on his body, has a great deal on his feet, making him look as if he wore pantalets, little short pantalets badly frayed out at the bottoms.

San Diego is one of the great new cities of Southern California. It lies within only a few minutes' ride of Mexico. There is a pretty little Mexican town on the line between Mexico and California—Tia Juana—pronounced Te Wanna. Translated, the name means "Aunt Jane." In the center of one of the streets stands a great gray stone monument, set there by the government to mark the line between the United States and Mexico.

To the south, several hundred miles distant, stretches the long Sea of Cortez, as the conquerors of ancient Mexico once called the Gulf of California. Beyond the Sea of Cortez is the long and rock-bound reach of the west coast of Mexico. Then a group of little Central American republics; then Colombia, Peru and so on, till at last Patagonia points away like a huge giant's finger straight toward the South Pole.

But I must bear in mind that I set out in this story to tell you about "The Bear-Slayer of San Diego," and the South Pole is a long way from the subject in hand.

I have spoken of San Diego as one of the great new cities, and great it is, but altogether new it certainly is not, for it was founded by a Spanish missionary, known as Father Junipero, more than one hundred years ago.

These old Spanish missionaries were great men in their day; brave, patient and very self-sacrificing in their attempts to settle the wild countries and civilize the Indians.

This Father Junipero walked all the way from the City of Mexico to San Diego, although he was more than fifty years old; and finally, after he had spent nearly a quarter of a century in founding missions up and down the coast of California, he walked all the way back to Mexico, where he died.

When it is added that he was a lame man, that he was more than threescore and ten years of age, and that he traveled all the distance on this last journey on foot and alone, with neither arms nor provisions, trusting himself entirely to Providence, one can hardly fail to remember his name and speak it with respect.

This new city, San Diego, with its most salubrious clime, is set all over and about with waving green palms, with golden oranges, red pomegranates, great heavy bunches of green and golden bananas, and silver-laden olive orchards. The leaf of the olive is of the same soft gray as the breast of the dove. As if the dove and the olive branch had in some sort kept companionship ever since the days of the deluge.

San Diego is nearly ten miles broad, with its base resting against the warm, still waters of the Pacific Ocean. The most populous part of the city is to the south, toward Mexico. Then comes the middle part of San Diego City. This is called "the old town," and here it was that Father Junipero planted some palm trees that stand to this day—so tall that they almost seem to be dusting the stars with their splendid plumes.

Here also you see a great many old adobe houses in ruins, old forts, churches, fortresses, barracks, built by the Mexicans nearly a century ago, when Spain possessed California, and her gaudy banner floated from Oregon to the Isthmus of Darien.

The first old mission is a little farther on up the coast, and the new college, known as the San Diego College of Letters, is still farther on up the warm sea bank. San Francisco lies several hundred miles on up the coast beyond Los Angeles. Then comes Oregon, then Washington, one of the newest States, and then Canada, then Alaska, and at last the North Pole, which, by the way, is almost as far as the South Pole from my subject: The Bear-Slayer of San Diego.

He was a little Aztec Indian, brown as a berry, slim and slender, very silent, very polite and not at all strong.

It was said that he had Spanish blood in his veins, but it did not show through his tawny skin. It is to be conceded, however, that he had all the politeness and serene dignity of the proudest Spanish don in the land.

He was now, by the kind favor of those good ladies who had heard of his daring address in killing the bear with his knife, a student of the San Diego College of Letters, where there were several hundred other boys of all grades and ages, from almost all parts of the earth.

A good many boys came here from Boston and other eastern cities to escape the rigors of winter. I remember one boy in particular from Philadelphia. He was a small boy with a big nose, very bright and very brave. He was not a friend of the little Aztec Indian, the Bear-Slayer of San Diego. The name of this boy from Philadelphia was Peterson; the Boston boys called him Bill Peterson. His name, perhaps, was William P. Peterson; William Penn Peterson, most likely. But this is merely detail, and can make but little difference in the main facts of the case.

As I said before, these college grounds are on the outer edge of the city. The ocean shuts out the world on the west, but the huge chaparral hills roll in on the east, and out of these hills the jack-rabbits come down in perfect avalanches at night, and devour almost everything that grows.

Wolves howl from these hills of chaparral at night by hundreds, but they are only little bits of shaggy, gray coyotes and do little or no harm in comparison with the innumerable rabbits. For these big fellows, on their long, bent legs, and with ears like those of a donkey, can cut down with their teeth a young orchard almost in a single night.

The new college, of course, had new grounds, new bananas, oranges, olives, all things, indeed, that wealth and good taste could contribute in this warm, sweet soil. But the rabbits! You could not build a fence so high that they would not leap over it.

"They are a sort of Jumbo grasshopper," said the smart boy from Boston.

The head gardener of the college campus and environment grew desperate.

"Look here, sir," he said to the president, "these big-eared fellows are lazy and audacious things. Why can't they live up in the chaparral, as they did before we came here to plant trees and try to make the world beautiful? Now, either these jack-rabbits must go or we must go."

"Very well," answered the president. "Offer a reward for their ears and let the boys destroy them."

"How much reward can I offer?"

"Five cents apiece, I think, would do," answered the head of the college, as he passed on up the great stone steps to his study.

The gardener got the boys together that evening and said, "I will give you five cents apiece for the ears of these dreadful rabbits."

"That makes ten cents for each rabbit, for each rabbit has two ears!" shouted the smart boy from Boston.

Before the dumfounded gardener could protest, the boys had broken into shouts of enthusiasm, and were running away in squads and in couples to borrow, buy or beg firearms for their work.

The smart boy from Boston, however, with an eye to big profits and a long job, went straight to the express office, and sent all the way to the East for a costly and first-class shotgun.

The little brown Aztec Indian did nothing of that sort; he kept by himself, kept his own counsel, and so far as any of the boys could find out, paid no attention to the proffered reward for scalps.

Bill Peterson borrowed his older brother's gun and brought in two rabbits the next day. The Boston boy, with an eye wide open to future profits to himself, went with Peterson to the head gardener, and holding up first one dead jack-rabbit by the ear, and then the other, coolly and deliberately counted off four ears.

The gardener grudgingly counted out two dimes, and then, with a grunt of satisfaction, carried away the two big rabbits by their long hind legs.

As the weeks wore by, several other dead rabbits were reported, and despite the grumbling of the head gardener, the tumultuous and merry students had quite a revenue, and their hopes for the future were high, especially when that artillery should arrive from Boston!

Meantime, the little brown Aztec boy had done nothing at all. However, when Friday afternoon came, he earnestly begged, and finally obtained, leave to go down to his home at Tia Juana. He wanted very much to see his Mexican mother and his six little Mexican brothers, and his sixty, more or less, little Mexican cousins.

But lo! on Saturday morning, bright and early, back came the little Bear-Slayer, as he was called by the boys, and at his heels came toddling and tumbling not only his six half-naked little brown brothers, but dozens of his cousins.

Each carried a bundle on his back. These bundles were long, finely woven bird-nets, and these nets were made of the fiber of the misnamed century plant, the agave.

This queer looking line of barefooted, bareheaded, diminutive beings, headed by the silent little Aztec, hastily dispersed itself along the outer edge of the grounds next to the chaparral abode of the jack-rabbits, and then, while grave professors leaned from their windows, and a hundred curious white boys looked on, these little brown fellows fastened all their long bird-nets together, and stretched two wide wings out and up the hill.

Very quiet but very quick they were, and when all the nets had been unwound and stretched out in a great letter V far up the hill, it was seen that each brown boy had a long, heavy manzanita wood club in his hand.

Suddenly and silently as they had come they all disappeared up and over the hills beyond, and in the dense black chaparral.

Where had they gone and what did all this silent mystery mean? One, two, three hours! What had become of this strange little army of silent brown boys?

Another hour passed. Not a boy, not a sign, not a sound. What did it all mean?

Suddenly, down came a rabbit, jumping high in the air, his huge ears flapping forward and back, as if they had wilted in the hot sun.

Then another rabbit, then another! Then ten, twenty, forty, fifty, five hundred, a thousand, all jumping over each other and upon each other, and against the nets, with their long legs thrust through the meshes, and wriggling and struggling till the nets shook as in a gale.

Then came the long lines of half-naked brown boys tumbling down after them out of the brush, and striking right and left, up and down, with their clubs.

In less than ten minutes from the time they came out of the brush, the little fellows had laid down their clubs and were dragging the game together.

The grave professors shook their hats and handkerchiefs, and shouted with delight from their windows overhead, and all the white boys danced about, wild with excitement.

That is, all but one or two. The boy from Boston said savagely to the little Aztec, as he stood directing the counting of the ears, "You're a brigand! You're the black brigand of San Diego City, and I can whip you!"

The brigand said nothing, but kept on with his work.

In a little time the president and head gardener came forward, and roughly estimated that about one thousand of the pests had been destroyed. Then the kindly president went to the bank and brought out one hundred silver dollars, which he handed to the little Bear-Slayer of San Diego in a cotton handkerchief.

The poor, timid little fellow's lips quivered. He had never seen so much money in all his life. He held his head down in silence for a long time and seemed to be thinking hard. His half-naked little brothers and cousins grouped about and seemed to be waiting for a share of the money.

The boy's schoolmates also crowded around, just as boys will, but they did not want any of the silver, and I am sure that all, save only one or two, were very glad because of his good luck.

Finally, lifting up his head and looking about the crowd of his school-fellows, he said, "Now, look here; I want every one of you to take a dollar apiece, and I will take what is left." He laid the handkerchief that held the silver dollars down on the grass and spread it wide open.

Hastily but orderly, his schoolmates began to take up the silver, his own little brown fellows timidly holding back. Then one of the white boys who had hastily helped himself saw, after a time, that the bottom was almost reached, and, with the remark that he was half ashamed of himself for taking it, he quietly put his dollar back. Then all the others, fine, impulsive fellows who had hardly thought what they were about at first, did the same; and then the little brown boys came forward.

They kept coming and kept taking, till there was not very much but his handkerchief left. One of the professors then took a piece of gold from his pocket and gave it to the little Bear-Slayer. The boy was so glad that tears came into his eyes and he turned to go.

"See here! I'm sorry for what I said. Yes, I am. I ought to be ashamed, and I am ashamed."

It was the smart boy from Boston who had been looking on all this time, and who now came forward with his hand held out.

"See here!" he said. "I've got a forty-dollar shotgun to give away, and I want you to have it. Yes, I do. There's my hand on it. Take my hand, and you shall have the gun just as soon as it gets here."

The two shook hands, and the boys all shouted with delight; and on the very next Saturday one of these two boys went out hunting quail with a fine shotgun on his shoulder.

It was the silent little hero, The Bear-Slayer of San Diego.

XV.

ALASKAN AND POLAR BEAR.

"And round about the bleak North Pole
　　Glideth the lean, white bear."

Nearly forty years ago, when down from the Indian country to sell some skins in San Francisco, I saw a great commotion around a big ship in the bay, and was told that a Polar bear had been discovered floating on an iceberg in the Arctic, and had been taken alive by the ship's crew.

I went out in a boat, and on boarding the ship, just down from Alaska with a cargo of ice, I saw the most beautiful specimen of the bear family I ever beheld. A long body and neck, short legs, small head, cream-white and clean as snow, this enormous creature stood before us on the deck, as docile as a lamb. This is as near as ever I came to encountering the Polar bear, although I have lived in the Arctic and have more than one trophy of the bear family from the land of everlasting snows.

Bear are very plenty in Alaska and the Klondike country, and they are, perhaps, a bit more ferocious than in California, for I have seen more than one man hobbling about the Klondike mines on one leg, having lost the other in an argument with bear.

As a rule, the flesh is not good, here, in the salmon season, for the bear is in all lands a famous fisherman. He sits by the river and, while you may think he is asleep, he thrusts his paw deep down, and, quick as wink, he lands a huge salmon in his bunch of long, hooded claws.

A friend and I watched a bear fishing for hours on the Yukon, trying to learn his habits. I left my friend, finally, and went to camp to cook supper. Then, it seems, my friend shot him, for his skin, I think. Thinking the bear dead, he called to me and went up to the bear, knife in hand. But the bear rose up when he felt the knife, caught the man in his arms and they rolled in the river together. The poor man could not get away. When we recovered his body far down the river next day, the bear still held him in her arms. She was a long, slim cinnamon, said to be the most savage fighter in that region.

All the bear of the far north seem to me to have longer bodies and shorter legs than in other lands. The black bear (there are three kinds of them) are bow-legged, I think; at least they "toe in," walk as an Indian walks, and even step one foot over the other when taking their time on the trail. We cultivated the acquaintance of a black bear for some months, on the Klondike, in the winter of '97-'98, and had a good chance to learn his habits. He was a persistent robber and very cunning. He would eat anything he could get, which was not much, of course, and when he could not get anything thrown to him from a door he would go and tear down a stump and eat ants. I don't know why he did not hibernate, as other bears in that region do. He may have been a sort of crank. No one who knew about him, or who had been in camp long, would hurt him; but a crowd of strangers, passing up the trail near our Klondike cabin, saw him, and as he did not try to get away he was soon dead. He weighed 400 pounds, and they sold him where he lay for one dollar a pound.

I fell in with a famous bear-hunter, a few miles up from the mouth of the Klondike early in September, before the snow fell, and with him made a short hunt. He has wonderful bear sense. He has but one eye and but one side of a face, the rest of him having been knocked off by the slash of a bear's paw. He is known as Bear Bill.

The moss is very deep and thick and elastic in that region, so that no tracks are made except in a worn trail. But Bill saw where a bit of moss had been disturbed away up on a mountain side, and he sat right down and turned his one eye and all his bear sense to the solution of the mystery.

At last he decided that a bear had been gathering moss for a bed. Then he went close up under a cliff of rocks and in a few minutes was peering and pointing down into a sunken place in the earth. And behold, we could see the moss move! A bear had covered himself up and was waiting to be snowed under. Bill walked all around the spot, then took position on a higher place and shouted to the bear to come out. The bear did not move. Then he got me to throw some rocks. No response. Then Bill fired his Winchester down into the moss. In a second the big brown fellow was on his hind feet looking us full in the face and blinking his little black eyes as if trying to make us out. Bill dropped him at once, with a bullet in his brain.

I greatly regret that I never had the good fortune to encounter a Polar bear, so that I might be able to tell you more about him and his habits; for men of science and writers of books are not bear-hunters, as a rule, and so real information about this white robber-monk of the cold, blue north is meager indeed. But here is what the most eminent English authority says about the nature and habits of this one bear that I have not shaken hands with, or encountered in some sort of way on his native heath:

"The great white bear of the Arctic regions—the 'Nennok' of the Eskimo— is the largest as well as one of the best known of the whole family. It is a gigantic animal, often attaining a length of nearly nine feet and is proportionally strong and fierce. It is found over the whole of Greenland; but its numbers seem to be on the decrease. It is distinguished from other bears by its narrow head, its flat forehead in a line with its prolonged muzzle, its short ears and long neck. It is of a light, creamy color, rarely pure white, except when young, hence the Scottish whalers call it the 'brounie' and sometimes the 'farmer,' from its very agricultural appearance as it stalks leisurely over the furrowed fields of ice. Its principal food consists of seals, which it persecutes most indefatigably; but it is somewhat omniverous in its diet, and will often clear an islet of eider duck eggs in the course of a few hours. I once saw it watch a seal for half a day, the seal continually escaping, just as the bear was about putting his foot on it, at the atluk (or escape hole) in the ice. Finally, it tried to circumvent its prey in another maneuver. It swam off to a distance, and when the seal was again half asleep at its atluk, the bear swam under the ice, with a view to cut off its retreat. It failed, however, and the seal finally escaped. The rage of the animal was boundless; it moaned hideously, tossing the snow in the air, and at last trotted off in a most indignant state of mind.

"Being so fond of seal-flesh, the Polar bear often proves a great nuisance to sealhunters, whose occupation he naturally regards as a catering to his wants. He is also glad of the whale carcasses often found floating in the Arctic seas, and travelers have seen as many as twenty bears busily discussing the huge body of a dead whalebone whale.

"As the Polar bear is able to obtain food all through the Arctic winter, there is not the same necessity, as in the case of the vegetable-eating bears, for hibernating. In fact, the males and young females roam about through the whole winter, and only the older females retire for the season. These—according to the Eskimo account, quoted by Captain Lyon—are very fat at the commencement of winter, and on the first fall of snow lie down and allow themselves to be covered, or else dig a cave in a drift, and then go to sleep until the spring, when the cubs are born. By this time the animal's heat has melted the snow for a considerable distance, so that there is plenty of room for the young ones, who tumble about at their ease and get fat at the expense of their parent, who, after her long abstinence, becomes gradually very thin and weak. The whole family leave their abode of snow when the sun is strong enough to partially melt its roof.

"The Polar bear is regularly hunted with dogs by the Eskimo. The following extract gives an account of their mode of procedure:

"Let us suppose a bear scented out at the base of an iceberg. The Eskimo examines the track with sagacious care, to determine its age and direction, and the speed with which the animal was moving when he passed along. The dogs are set upon the trail, and the hunter courses over the ice in silence. As he turns the angle of the berg his game is in view before him, stalking along, probably, with quiet march, sometimes snuffing the air suspiciously, but making, nevertheless, for a nest of broken hummocks. The dogs spring forward, opening a wild, wolfish yell, the driver shrieking 'Nannook! Nannook!' and all straining every nerve in pursuit.

Pressed more severely, the bear stands at bay.

"The bear rises on his haunches, then starts off at full speed. The hunter, as he runs, leaning over his sledge, seizes the traces of a couple of his dogs and liberates them from their burthen. It is the work of a minute, for the motion is not checked, and the remaining dogs rush on with apparent ease.

"Now, pressed more severely, the bear makes for an iceberg, and stands at bay, while his two foremost pursuers halt at a short distance and await the arrival of the hunter. At this moment the whole pack are liberated; the hunter grasps his lance, and, tumbling through the snow and ice, prepares for the encounter.

"If there be two hunters, the bear is killed easily; for one makes a feint of thrusting the spear at the right side, and, as the animal turns with his arms toward the threatened attack, the left is unprotected and receives the death wound."

XVI.

MONNEHAN, THE GREAT BEAR-HUNTER OF OREGON.

He wore a tall silk hat, the first one I had ever seen, not at all the equipment of "a mighty hunter before the Lord;" but Phineas Monnehan, Esq., late of some castle (I forget the name now), County of Cork, Ireland, would have been quite another personage with another sort of hat. And mighty pretension made he to great estates and titles at home, but greatest of all his claims was that of "a mighty hunter."

Clearly he had been simply a schoolmaster at home, and had picked up all his knowledge of wild beasts from books. He had very impressive manners and had come to Oregon with an eye to political promotion, for he more than once hinted to my quiet Quaker father, on whose hospitality he had fastened himself, that he would not at all dislike going to Congress, and would even consent to act as Governor of this far-off and half-savage land known as Oregon. But, as observed a time or two before, Monnehan most of all things desired the name and the renown, like Nimrod, the builder of Babylon, of a "mighty hunter."

He had brought no firearms with him, nor was my father at all fond of guns, but finally we three little boys, my brother John, two years older than I, my brother James, two years younger, and myself, had a gun between us. So with this gun, Monnehan, under his tall hat, a pipe in his teeth and a tremendously heavy stick in his left hand would wander about under the oaks, not too far away from the house, all the working hours of the day. Not that he ever killed anything. In truth, I do not now recall that he ever once fired off the gun. But he got away from work, all the same, and a mighty hunter was Monnehan.

He carried this club and kept it swinging and sweeping in a semi-circle along before him all the time because of the incredible number of rattlesnakes that infested our portion of Oregon in those early days. I shall never forget the terror in this brave stranger's face when he first found out that all the grass on all our grounds was literally alive with snakes. But he had found a good place to stay, and he was not going to be driven out by snakes.

You see, we lived next to a mountain or steep stony hill known as Rattlesnake Butte, and in the ledges of limestone rock here the rattlesnakes hibernated by thousands. In the spring they would crawl out of the cracks in the cliffs, and that was the beginning of the end of rattlesnakes in Oregon. It was awful!

But he had a neighbor by the name of Wilkins, an old man now, and a recent candidate for Governor of Oregon, who was equal to the occasion. He sent back to the States and had some black, bristly, razor-backed hogs brought out to Oregon. These hogs ate the rattlesnakes. But we must get on with the bear story; for this man Monnehan, who came to us the year the black, razor-backed hogs came, was, as I may have said before, "a mighty hunter."

The great high hills back of our house, black and wild and woody, were full of bear. There were several kinds of bear there in those days.

"How big is this ere brown bear, Squire?" asked Monnehan.

"Well," answered my father, "almost as big as a small sawmill when in active operation."

"Oi think Oi'll confine me operations, for this hunting sayson, to the smaller spacies o' bear," said Mr. Monnehan, as he arose with a thoughtful face and laid his pipe on the mantel-piece.

A few mornings later you would have thought, on looking at our porch, that a very large negro from a very muddy place had been walking bare-footed up and down the length of it. This was not a big bear by the sign, only a small black cub; but we got the gun out, cleaned and loaded it, and by high noon we three little boys, my father and Monnehan, the mighty hunter, were on the track of that little black bear. We had gone back up the narrow canyon with its one little clump of dense woods that lay back of our house and reached up toward the big black hills.

Monnehan took the gun and his big club and went along up and around above the edge of the brush. My father took the pitchfork and my younger brother James kept on the ridge above the brush on the other side of the canyon, while my older brother John and myself were directed to come on a little later, after Mr. Monnehan had got himself in position to do his deadly work, and, if possible, drive the terrible beast within range of his fatal rifle.

Slowly and cautiously my brother and I came on, beating the brush and the tall rye grass. As we advanced up the canyon, Mr. Monnehan was dimly visible on the high ridge to the right, and father now and then was to be seen with little brother and his pitchfork to the left. Suddenly there was such a shout as almost shook the walls of the canyon about our ears. It was the voice of Monnehan calling from the high ridge close above the clump of dense wood; and it was a wild and a desperate and a continuous howl, too. At last we could make out these words:

"Oi've thrade the bear! Oi've thrade the bear! Oi've thrade the bear!"

Down the steep walls came father like an avalanche, trailing his pitchfork in one hand and half dragging little brother James with the other.

"Run, boys, run! right up the hill! He's got him treed, he's got him treed! Keep around the bush and go right up the hill, fast as you can. He's got him treed, he's got him treed! Hurrah for Monnehan, at last! He's got him treed, he's got him treed!"

Out of breath from running, my father sat down at the foot of the steep wall of the canyon below Monnehan and we boys clambered on up the grassy slope like goats.

Meantime, Monnehan kept shouting wildly and fearfully as before. Such lungs as Monnehan had! A mighty hunter was Monnehan. At last we got on the ridge up among the scattering and storm-bent and low-boughed oaks; breathless and nearly dead from exhaustion.

"Here, byes, here!"

We looked up the hill a little ahead of us from where the voice came, and there, straddled across the leaning bough of a broad oak tree hung Monnehan, the mighty hunter. His hat was on the ground underneath him, his club was still in his daring hand, but his gun was in the grass a hundred yards away.

"Here, boys, right up here. Come up here an' get a look at 'im! Thot's vaght Oi got up 'ere fur, to get a good look at 'im! Right up now, byes, an' get a good look at 'im! Look out fur me hat there!"

My brother hastily ran and got and handed me the gun and instantly was up the tree along with Monnehan, peering forward and back, left and right, everywhere. But no sign, no sound or scent of any bear anywhere.

By this time my father had arrived with his pitchfork and a very tired little boy. He sat down on the grass, and, wearily wiping his forehead, he said to Monnehan,

"Mr. Monnehan, how big was the bear that you saw?"

"Well, now, Squire, upon the sowl o' me, he was fully the size of a very extraordinary black dog," answered Mr. Monnehan, as he descended and came and stood close to my father, as if to defend him with his club. Father rose soon after and, with just the least tinge of impatience and vexation in his voice, said to brother John and me,

"Boys, go up and around the thicket with your gun and beat the bush down the canyon as you come down. Mr. Monnehan and I will drop down to the bottom of the canyon here between the woods and the house and catch him as he comes out."

Brother and I were greatly cheered at this; for it was evident that father had faith that we would find the bear yet. And believing that the fun was not over, we, tired as we were, bounded forward and on and up and around the head of the canyon with swift feet and beating hearts. Here we separated, and each taking a half of the dense copse of wood and keeping within hailing distance, we hastily descended through the steep tangle of grapevine, wild hops, wild gourdvines and all sorts of things, shouting and yelling as we went. But no bear or sign of bear as yet.

We were near the edge of the brush. I could see, from a little naked hillock in the copse where I paused to take breath, my father with his pitchfork standing close to the cow path below the brush, while a little further away and a little closer to the house stood Mr. Monnehan, club in hand and ready for the raging bear.

Suddenly I heard the brush break and crackle over in the direction of my brother. I dropped on my knee and cocked my gun. I got a glimpse of something black tearing through the brush like a streak, but did not fire.

Then I heard my brother shout, and I thought I heard him laugh, too. Just then there burst out of the thicket and on past my father and his pitchfork a little black, razor-backed sow, followed by five black, squealing pigs! Monnehan's bear!

XVII.

THE BEAR "MONARCH."
HOW HE WAS CAPTURED.

Much having been said about bears of late, a young Californian of great fortune and enterprise resolved to set some questions at rest, and, quite regardless of cost or consequences, sent into the mountains for a live grizzly. The details of his capture, the plain story of the long, wild quest, the courage, the cunning, the final submission of the monster, and then the last bulletin about his health, habits and all that, make so instructive and pleasing a narrative that I have asked for permission to add it to my own stories. The bear described is at present in our San Francisco Zoo, a fine and greatly admired monarch.

"Are there any true grizzly bears in California?"

"Undoubtedly there are."

"I don't know about it. I have a great deal of doubt. Where are they?"

"In the Sierra Madre, in Touloumne Canyon, in Siskiyou County and probably in many other mountain districts."

"That may be so, but nobody can find them. Now, do you think you could find them?"

"I think I could if I should try."

"Would you undertake to get a genuine grizzly in this State?"

"Yes, if you want one. How will you have him—dead or alive?"

"Alive."

This conversation was held last May between the proprietor of the Examiner and special reporter Allen Kelly.

A week ago Kelly brought home an enormous grizzly bear, lodged the animal temporarily in one of the cages in Woodward's Gardens and reported to the editor that he had finished that assignment.

The following is his account of the hunt and capture.

The Examiner expedition began the search for a grizzly early in June, starting from Santa Paula and striking into the mountains at Tar Creek, where the Sespe oil wells are bored. The Examiner correspondent detailed to catch a bear was accompanied by De Moss Bowers of Ventura, who was moved by love of adventure to offer his assistance.

During the first part of the trip the party numbered five persons, including Dad Coffman, a spry old gentleman of seventy-two years, who was out for the benefit of his health, a packer and guide, and a person from Santa Paula called "Doc," who was loaded to the muzzle with misinformation and inspired with the notion that it was legitimate to plunder the expedition because the Examiner had plenty of money. The packer was "Doc's" son, a good man to work, but unfortunately afflicted with similar hallucinations. The expedition was plundered because these persons were trusted on the recommendation of a gentleman who ought to have known better.

At Tar Creek the correspondent was told that the Stone Corral bear, a somewhat noted grizzly that had killed his man, had been recently on Squaw Flat, and had prowled about an old cabin at night, sorting over the garbage heap and pile of tin cans at the door, but when the expedition passed the cabin no fresh sign was found, and the tracks on Squaw Flat were at least a week old.

The first camp was in a clump of chincapin brush at Stone Corral. There were bear tracks in the soft ground at the edge of the creek, which induced the hunters to spend two days in prospecting that part of the country. One of the proposed plans for capturing the bear was to run him out of the rocks and brush to some reasonably open bit of country like Squaw Flat or one of the small level patches near camp and lasso him, but the impracticable nature of that scheme was soon demonstrated. On the next day after making camp the Examiner's own bear catcher went out on a nervous black horse called "Nig" to find out where the Stone Corral bear was spending the summer and incidentally to get some venison. The Stone Corral bear was there or thereabouts beyond any doubt. He ran the correspondent out of the brush and showed a perverse disposition to do all the hunting himself. "Nig" would not stand to let his rider take a shot, but when the bear gave notice of his presence by growling and smashing down the brush twenty yards away, he wheeled and bolted towards camp. Near the camp Dad was found rounding up the other horses, who had just been scared from their pasturage by another wandering bear. It was clear that not a horse in the outfit could be ridden to within roping distance of a bear, and it is doubtful if three horses fit for such a job could be found in the country. Some years ago the ranchmen and vaqueros frequently caught bears with a rope, but even then it was difficult to train horses to the work, and only one horse out of a hundred could be cured of his instinctive dread of a grizzly.

It was clear also that there were some defects in the plan of driving the Stone Corral bear out of the brush, chief of which was the bear's inconsiderate desire to do the driving himself. As the hunting would have to be done afoot, the prospects incident to an attempt to round up a big grizzly among the rocks and chaparral were not peculiarly alluring. Trapping was the only other method that could be suggested, but the absence of any heavy timber would make that difficult.

The Stone Corral is a singular arrangement of huge sandstone ledges on the slope of a mountain, forming a rough inclosure about a quarter of a mile wide and three or four times as long. The country is very rugged and broken for miles around, and except along the creek and on the trail a horse cannot be ridden through it. The problem of how to catch a bear in such a place was not solved, because the bear cut short its consideration by marching past the camp and lumbering down the creek bed toward the Alder Creek Canyon and the Sespe country. The correspondent stood upon the sandstone ledge as he went by, and yelled at him, but he did not quicken his pace.

When it became evident that the bear was bound for the Sespe, the horses were saddled. Balaam the Burro was concealed under a mountainous pack, and the march was resumed over the Alder Creek trail to the deep gorge through which the Sespe River runs. The man who made the Alder Creek trail was not born to build roads. He laid it out right over the top of a high and steep mountain, when by making a slight detour, he could have avoided a difficult and unnecessary climb. In the broiling hot sun of a breezeless day in June, the march over the mountains was hard on men and horses, and the pace was necessarily slow.

The heat coaxed the rattlesnakes out of their holes, and the angry hum of their rattles was an almost incessant accompaniment to the hoof beats of the horses. Where the trail wound along a steep slope, affording but slight foothold for an animal, a more than usually strenuous and insistent singing of a snake, disturbed from his sunny siesta, caused Balaam to jump aside. Balaam avoided the snake, but he lost his balance and rolled down the slope, heels in the air and pack underneath. The acrobatic feats achieved by Balaam in his struggles to regain his footing were watched by an admiring and solicitous audience, and when he cleverly took advantage of the slight obstruction offered by a manzanita bush, and got safely upon his feet, he was loudly applauded. The deep solicitude of the party for the safety of Balaam and his pack was accounted for when he scrambled back to the trail and gravely walked up to the packer to have his pack straightened. Every man anxiously felt of the pack, and heaved a sigh of relief. The bottles containing O. P. S., antidote for snake bite, were not broken, but it was a narrow escape.

"Great Beeswax!" said the Doctor, "suppose those bottles had been smashed and then some one of us should go to work and bite himself with a snake! Wouldn't that be a fix?"

"Dogdurn if it don't make my blood run cold to think of it," said Dad.

Everybody's blood seemed to be congealing, and as the pack was loose and the antidote accessible, an ounce of prevention was administered to each man, and Balaam was rewarded for his timely agility with a handful of sugar.

No more accidents occurred, and late in the afternoon the cavalcade slid, coasted and scrambled down the last steep hill into the Sespe Canyon, where a camp was made under an immense oak beside a deep, rocky pool. That evening, around the camp-fire, some strange bear stories were evolved from either the memories or imagination of the hunters.

In the morning the search for bear signs was resumed and prosecuted until noon without success. Dad was lured by the swarms of trout in the stream, and went fishing. Dad is not a scientific fly fisherman. His favorite method is to select a shady nook on the bank, sit down with his back against a rock, tie a sinker to a large and gaudy fly, and angle on the bottom for the biggest trout he can see. He generally carries a book in his pocket, and when the trout remains unresponsive to the allurements of the gaudy fly, he fastens his rod to a bush and reads until he falls asleep.

In the afternoon one of the party went out over a long, brushy ridge, and the correspondent pushed on down the gorge in search of bear signs. All the bear tracks led up toward the Hot Springs Canyon, indicating that the grizzlies had begun their annual migration to the Alamo, Frazier and Pine mountains, where large bands of sheep are herded through the summer. Some of the tracks were large and fresh, and a person might come upon a bear at any time in the bottom of the canyon. Preparations were made for following the bears and directions given for an early start in the morning. The Doctor recollected that he had important business in Santa Paula that required his immediate attention, and he wouldn't have time to follow the grizzlies through the rugged passes of the mountains. Accordingly, he and Dad decided to remain in the Sespe camp a day or two, enjoy the fishing, and then return to Santa Paula, and the bear hunting party that saddled up and struck out on the trail of the grizzly in the morning was reduced to three.

The trail led through the Hot Springs Canyon, where boiling hot sulphur water flows out of the ground in a stream large enough to sensibly affect the temperature of the Sespe River, into which it runs. This canyon was formerly a beautiful camping spot, and was resorted to by many persons who believed that bathing in sulphur water would restore their health, but about three years ago a cloudburst uprooted all the trees and converted the green cienaga into a rocky desolate flat, as barren and unattractive as the sharp, treeless peaks surrounding the canyon. A few mountain sheep inhabit the mountains about the Hot Springs, and occasionally one is seen standing upon some high and inaccessible cliff, but it is very seldom that a hunter succeeds in getting a pair of big horns.

The next camp was on the Piru Creek, where it runs through the Mutaw ranch. One of the most promising mining districts in this part of the State takes its name from the Piru, and in years gone by a great deal of gold was taken from the diggings along the stream. One of the most successful miners was Mike Brannan, whose cabins and mining appliances lie unused and decaying about six miles from the place where the expedition camped.

From the camp on the Mutaw the expedition followed Piru Creek down to Lockwood, and the latter up to the divide between Lockwood Valley and the Cuddy ranch at the foot of Mount Pinos, called Sawmill Mountain by the settlers. The mountain is about 10,000 feet high, and is covered with heavy pine timber. Ever since Haggin & Carr's sheep have been on the mountain, the bears from forty miles around have made annual marauding expeditions, and kept the herders on the jump all the summer. The first band of sheep and the Examiner expedition arrived at the old Sawmill simultaneously this year, and the Basque who was herding the band, having a very lively sense of the danger of his situation, pitched his tent close to the camp, where he would be under the protection of three rifles. The Basque had never been on the mountain before, but he had heard about the bears and their audacious raids, and he was not at all enamored of his job. When the campfires were started, and the forest became an enclosing wall of gloom, behind which lurked all the mysteries and menaces of the mountains, the Basque came shyly into camp, bringing a shoulder of mutton with which to establish friendly relations, and under the mellowing influence of a glass of something hot he became confidential and as communicative as his broken jargon of French and California Spanish would permit.

He had come to the mountain reluctantly, and having been told about the herder whose hand was torn off by a grizzly last year, he was still more unwilling to remain. He would stay as long as the Examiner party remained near him, but when the hunters went away he proposed to quit and hasten back to the plains, where he would have nothing worse than the coyotes to encounter. Every night after that, so long as the hunters were in that camp, the Basque came and sat at the fire until bedtime, talking about *los osos*, and when the grass and water gave out and the expedition was obliged to move camp about two miles, the gentle shepherd packed his blankets over the trail to Bakersfield, leaving his flock in the care of a leathery skinned bear-hardened Mexican.

The bears were later this year than usual in coming to the mountain, probably because the warm weather was longer delayed, and for many days the hunters scanned the trails in the canyons in vain for the footprints of grizzlies. The first indication of their arrival was given in a somewhat startling way to the correspondent one evening as he was slowly toiling through a deep, rocky ravine back to camp, after a weary tramp over the foothills of the big mountain.

The sun had set and the bottom of the ravine was dark as night. The belated searcher for bear signs skirted a dense willow thicket, and brushed against the bushes with his elbow. "Woof! Woof!" snorted a bear within ten feet of him, invisible in the thicket. His heart thumped and his rifle lock clicked, together, and which sound was the louder he could not tell. For a few seconds he stood at the edge of the thicket with his rifle ready, expecting the rush of the bear, but the animal was not in a warlike mood and did not rush, and the hunter cautiously backed away about twenty yards up the steep side of the ravine. The cracking of brush indicated that bruin was moving in the thicket, but nothing could be seen in the gathering gloom. Two or three large rocks rolled down into the willows started the bear out on a run and he could be heard crashing his way down the ravine and splashing into the pools as he went. The remainder of the journey back to camp was made through the open pine forest on the top of the mountain.

Superintendent McCullough, who has charge of Haggin & Carr's sheep camps on Pinos Mountain, stopped at the Examiner camp when he made his inspecting tours, and consultations were held with him about the bears. From the reports given him by the herders he judged that only the bears that lived on the mountain were prowling about, and that the invading army had not arrived from the Alamo and the Sespe region. A large cinnamon bear had walked into one camp about ten miles distant and killed two sheep in daylight, but the grizzlies had not begun to eat mutton. In July or August there would be bears enough to keep a man busy shinning up trees. Last year, he said, there were at least forty bears on the mountain, and they visited some of the sheep camps every night. Sometimes two or three bears would raid a camp, tree the herder and kill several sheep. The herders were not expected to fight bears or attempt to drive them away, and the owners reckoned upon the loss of several hundred sheep every summer.

Shortly before the first of July the camp was moved to Seymore Spring, about two miles from the mill, where good water and feed were plenty, and search for bear sign was continued. Every day some deep gorge or rocky ravine was visited and thoroughly hunted, and a deer was killed occasionally, but no sign of bears was found until the 3d of July, when the tracks of a very large grizzly were discovered crossing a ridge between the Lockwood Valley and the

Seymour. The tracks were followed across the Seymour Valley to a spur of the mountain between the mill ravine and a deep canyon to the westward.

Camp was moved to a green cienaga at the head of the latter, which was christened Bear Canyon, and the building of a trap was begun near the mouth—about half a mile from camp. Three large pine trees served as corner posts for a pen built of twenty-inch logs, "gained" at the corners and fastened together with stout oak pins. The pen was about twelve feet long, four feet high and five feet wide inside, and the door was made of pine logs sunk into the ground and wedged and pinned securely. A door of four-inch planks, so heavy that it required three men to raise it, was set in front, between oak guides pinned vertically to the trees and suspended by a rope running over a pulley and back to a trigger that engaged with a pivoted stick of oak, to which the bait was to be fastened. Five days were consumed in the construction of the trap, and while the work was going on a bear visited the camp at night and stampeded all the saddle and pack animals out of the canyon.

A German prospector named Sparkuhle, who was staying temporarily in the camp, was cured of a severe case of skepticism that night. Sparkuhle believed nothing that he could not see, and he declared, with exasperating iteration, "I believe there don't vas any bears in der gountry. I look for 'em every day, thinking perhaps might I could see one, but I don't could see any." And every night before he turned in, Sparkuhle said: "Vell, might did a bear come tonight. I wish I could see one, but I think there don't vas any bears at all."

Sparkuhle scorned the shelter of the bough shed, under which the Examiner outfit slept, and spread his blankets on top of a bank about six feet above a rocky shelf that was used as a pantry and kitchen. His only weapon was his pick, and he was not afraid of being disturbed by any prowling animal.

It was about midnight when the camp was alarmed by the snorting of the horses and the clatter of hoofs galloping down the canyon, but before the cause of the disturbance could be learned a yell of surprise came from Sparkuhle, followed by a crash and a terrible clatter among the pots and pans below the bank. In another moment Sparkuhle ran into the camp and began to tell excitedly what had happened to him. He was so intensely interested in his story that he paid no attention to a three-tined fork that was sticking in him just below the end of his back. He said he was awakened by the noise in camp, and looking up thought he saw the burro standing over him. Seizing his pillow he made a swipe at the animal, and said, "Get away, Balaam!" and then the supposed burro hit him a clip and knocked him spinning over the edge of the bank, but the blow did no further damage because Sparkuhle was rolled up in half a dozen blankets. The noise of his arrival among the tinware alarmed the bear and when the party got out with lights and guns he was out of sight. Sparkuhle slept in the cabin after that.

Two days later the big bear went into a sheep camp near the mill, while the herder was cooking supper, stampeded the sheep right over the fire, caught one and killed it, and sat down within thirty yards of the herder and leisurely gorged himself with mutton. The Mexican herder described him as "grande" and "muy blanco" and said he was as tall as a mule. On the following day at noon the same bear went into another sheep camp about three miles from the mill, and stole a freshly killed sheep, which the herder had hung up for his own use. Then he suddenly ceased his raids and disappeared and for the next three weeks the mountain seemed to be deserted by the bears.

The herders had put strychnine into the carcasses of several sheep that had died of eating poisonous weeds, and McCullough thought the bears must have eaten the poisoned mutton and become sick. It requires a strong dose of strychnine to kill a grizzly, and frequently the bears get only enough to make them ill and send them into temporary retirement in some dark gorge.

But while the bears were away the mountain lions and panthers managed to keep things from becoming dull. They came into camp several times and made the canyon ring with their yowling, but they always kept brush between themselves and the fire-light, and it was impossible to get a shot at them. Their raids became so annoying that two hounds were procured and brought into camp; after that the nightprowling beasts kept at a respectful distance. Being unable to steal any more provisions from the Examiner outfit, the lions turned their attention to the sheep camps. One night a lion sneaked up through a willow thicket to the nearest sheep camp and killed three sheep. He was a dainty lion, evidently, as he only cut the throats of the sheep and drank their blood and did not eat any mutton. The same lion followed the scent of a carcass that had been dragged to the bear trap for bait, but he stopped twenty yards from the trap, and went away, not caring to risk his neck by going into any such contrivance.

Wherever bait was dragged over the mountain, and it was dragged many miles for the purpose of enticing bear to the trap, the lions followed the trail, but they would not go into the trap. Still it is not safe to generalize from this fact and assume that the cougar or mountain lion never will go into a trap, for he is a most erratic and uncertain beast. Sometimes he is an arrant coward, and again he is as bold as a genuine lion. Generally a dog will keep cougars away from a camp or house, but once in a while the cougar hunts the dog and kills him.

One afternoon a cougar jumped into Joe Dye's dooryard at his ranch on the Sespe, picked up Joe's baby and sprang over the fence with it. Joe seized his rifle and shot the animal as it ran, and when the cougar felt the sting of the bullet he dropped the baby and ran up the mountain. He had seized the baby's clothes only, and the little one was not hurt. The next night the cougar returned, captured Joe's hound, carried it into the mountains and killed it.

On the 1st of August, the report reached camp that the bears were having a picnic on the Mutaw ranch and were killing hogs by the score. John F. Cuddy's sons, the best vaqueros and bronco-riders in this part of the country, offered to go over to the Mutaw with the correspondent and lasso a bear if one could be found on open ground; accordingly, the party saddled up and took the trail up the Piru, arriving at the Mutaw meadows late in the night, after a rough ride of twenty miles.

In the morning Mr. Taylor, one of the owners of the ranch, was found skinning a grizzly that had eaten strychnine in pork during the night. Mr. Taylor had put poison out all over the ranch and the prospect of catching a live bear seemed dubious, but all the poisoned meat that could be found was buried at once, and Bowers and the correspondent began building a trap to catch a bear that had been making twelve-inch tracks around the cabins. The Cuddy boys rode about looking for bear, and one of them lassoed an eagle that had waterlogged himself and was sitting stupidly on a rock by the creek. The bird measured nine feet across the wings. Messrs. Louis and Taylor, owners of the Mutaw, received the party hospitably and assisted in the work of preparing the trap. But Mr. Taylor forgot where he had put some of his poison, and in forty-eight hours all the dogs in the place, including the Examiner's two hounds, were stiffened out and turned up their toes. Chopping off their tails and pouring sweet oil down their throats did not restore them.

No chance to lasso a bear presented itself, and as soon as the trap was completed and baited with two live pigs the party returned to Pine Mountain.

At last it became evident that the bears on Mount Pinos could not be enticed into a trap while they had their pick and choice of the thousands of sheep that grazed on the mountain. They preferred to do their own butchering and would not touch mutton that was killed for them by anybody else. A cougar raided a camp one night, sprang upon the sheep from a willow thicket and killed three within twenty yards of the sleeping herder. The fastidious cougar cut their throats, sucked their blood and left their carcasses at the edge of the thicket without eating the meat. But the bears would not touch what the cougar left.

Shortly after this the herders reported that the bears were avoiding the sheep and passing around the bands without making an attack.

Apparently bruin had made a miscalculation in his calendar and was keeping Lent in the wrong season, but his erratic conduct was explained when some of the herders admitted that they had put strychnine into several carcasses.

Some of the bears had got doses of poison large enough to make them mortally unwell, but had survived and sworn off eating mutton. They disappeared from the vicinity of the camps and grazing ground, and went into solitary confinement in remote and deep gorges, where nobody but a lunatic would follow them.

The result of many weeks' hard work on Mount Pinos was the acquirement of some knowledge of the nature and eccentricities of Ursus ferox, which was glibly imparted by Tom, Dick and Harry, who assumed that the mere fact of their having lived near the mountains qualified them to speak as authorities on the habits of bears.

One inspired idiot declared that the best way to catch a grizzly was to give him atropia, which would make him blind for a day or two, and lead him along like a tame calf. This genius was so enamored of his great discovery that he went about the country telling everybody that the Examiner man was going to catch a grizzly with atropia, and that he (the aforesaid lunatic) was the inventor of the scheme and general boss of the outfit.

"A bear will do this," said one. "He will do so and so," said another, and "you just do that and he'll go right into the trap," said a dozen more. Everybody seemed to be loaded to the guards with an assorted cargo of general ignorance about bears, which they were anxious to discharge upon the Examiner expedition, but not one man in the whole lot ever caught a grizzly, and very few ever saw one.

As a matter of fact, determined by experience and observation, a grizzly will do none of the things laid down as rules of conduct for him by the wise men of the mountains, but will do pretty much as he pleases, and act as his individual whim or desire moves him. It is a mistake to generalize about bears from the actions of one of the species. One bear will be bold and inquisitive, and will walk right into a camp to gratify his curiosity, while another will carefully avoid man and all his works.

The predictions of an ursine invasion of Mount Pinos were not fulfilled and when it became clear that the few grizzlies in the neighborhood were too timid and wary to be caught, the expedition struck camp and moved on, leaving the traps set for luck.

Considerable annoyance was caused by a discharged mule-packer, who carried away tools required in trap building, and embezzled quite a sum of money. The fellow had attempted to impose upon the correspondent by whittling out pine-bark models of bear's feet, with which to make tracks around the trap; and had proposed various swindling jobs to others of the party, explaining that the "Examiner was rich and they might as well get a hack at the money." He had opened and read letters intrusted to him for mailing, and had proved himself generally a faithless scamp and an unconscionable liar. A written demand upon him, for restitution of his plunder, elicited only a coarse and abusive letter, but there was no time to waste in prosecuting the fellow and he was left in the enjoyment of his booty and in such satisfaction as the rascal mind of him could derive from the fact that he had succeeded in robbing his employer.

The big bear on the Mutaw never came near the trap built for his special accommodation, notwithstanding the confident assurances of the bear experts on the ranch that he was sure to show up within forty-eight hours. For two months after the poisoning of his campanero no signs of the large grizzly were seen anywhere near the Mutaw, and the hogs roamed about the hills unmolested.

After leaving Mount Pinos the expedition built several traps in the mountains near trails frequented by bears. An old grizzly that lived among the unsurveyed and unnamed peaks between Castac Lake and the Liebra Mountain absorbed the attention of the hunters for some time. He was an audacious marauder and killed his beef almost within sight of the camp-fire. Often at night a cow or steer could be heard bellowing in terror, and in the morning a freshly killed animal would be found in some hollow not far away, bearing marks of bear's claws. Whitened bones scattered all over the hills showed that the bear had been the boss butcher of General Beal's ranch for a long time. His average allowance of beef appeared to be about two steers a week, but he usually ate only half a carcass, leaving the rest to the coyotes and vultures.

One morning Bowers returned from a hunt for the horses, two of which had been struck and slightly wounded by the bear a few nights before, and had run away, and reported the discovery of a dead steer within 150 yards of an unfinished trap, about a quarter of a mile from camp. The animal appeared to have been killed two nights before, and the bear had made but one meal off the carcass. As he might be expected to return that night, all haste was made to finish the trap. Bowers rode out to Gorman's Station to get some nails and honey, while the correspondent paid a visit to one of General Beal's old corrals and stole some planks to make a door. He packed the planks up the mountain, and was using the hammer and saw with great diligence and a tremendous amount of noise, when bruin sauntered down the ridge, looked curiously at him and calmly began eating an early supper, wholly indifferent to the noise of the hammer and the presence of the man.

It was nearly dark when Bowers rode up to the trap, his horse in a lather composed of equal parts of perspiration and honey, the latter having leaked profusely from the cans tied to the saddle. Tossing the nails to the correspondent, Bowers hastily dismounted and went afoot up the ridge toward the dead steer, intending to place a can of honey near it. In about a minute Bowers was seen running from the ridge in fifteen-foot jumps, and as he approached the trap he shouted: "The bear is there now!"

"Is that so?" said the correspondent. "I thought he had finished his supper and had gone away by this time."

Bowers had approached to within forty yards of the bear before seeing him, and the bear had merely raised his head, taken a look at the intruder and resumed his eating. As it had become too dark to drive nails, and there was no longer any reason for finishing the door that night, Bowers fetched the rifles from camp and the two men went up the ridge to take a better look at the bear. Had there been light enough to make the rifle sights visible, it would have been hard to resist the temptation of turning loose at the old fellow from behind a convenient log; but it was impossible to draw a bead on him, and it would have been sheer foolhardiness to shoot and take the chances of a fight in the dark with a wounded grizzly. Besides, if shot at and missed, the bear would probably not return, and all the chances of getting him into the trap would be lost. So the two sat on a log and watched the grizzly till the night came on thick and dark, when they returned to camp.

The trap was finished the next day, but a somewhat ludicrous accident destroyed its possibilities of usefulness, and made it quite certain that bruin would never be caught in it. Not expecting a visit from the bear, for at least two days, the correspondent went up to the ridge just before dark, made a rope fast to the remains of a steer, and dragged him down to the trap. Bowers had gone back to Ventura on business, and the correspondent was alone on the mountain; when he went into the trap to fix a can of honey upon the trigger, he placed a stick under the door, in such a way that if the door should fall he could use the stick as a lever to pry it up, and so avoid an experience like Dad Coffman's.

The precaution was well taken. While he was arranging the bait he heard snuffling and the movement of some animal outside. Supposing that some cow or perhaps the burro was wandering about, he paid no particular attention to the noise, but when the bait was arranged and he turned to go out he saw the muzzle of old bruin poked into the door and his eyes blinking curiously at the dark interior of the trap. Bruin had come down for a feast and had followed the trail of the steer's remains with unexpected promptness. He had scented the honey, which was more alluring than stale beef, and evidently was considering the propriety of entering the trap to get his supper, which might consist of honeycomb *au naturel*, with Examiner man on the side.

The man in the trap deemed it highly improper for the bear to intrude at that time, and quickly decided the etiquette of the case by kicking the trigger and letting the door fall with a dull thud plump upon the old grizzly's nose. A hundred and sixty pounds falling four feet is no laughing affair when it hits one on the nose, and bruin did not make light of it. He was pained and surprised, and he went away more in sorrow than in anger, judging from the tone of his expostulating grunts and snorts.

When the snorts of the bear died away in the distance, the correspondent pried up the door, crawled out and cautiously made his way through the dark woods to his lonely camp.

At this time there were six traps scattered through the mountains within a radius of sixty miles, all of them set and baited, and the more distant ones watched by men employed for that purpose. One of the traps was on a mountain that was not pastured by cattle, or sheep, and as there were no acorns in that part of the country, the bears had to rustle for a living and were unable to withstand the temptation offered by quarters of beef judiciously exposed to their raids.

The bait scattered around this trap was discovered by four bears, but for some time they regarded it with suspicion, and were afraid to touch it, possibly because they detected the scent of man near it. Gradually they became accustomed to it and the signs of man's presence, and then they began to quarrel over the meat, as was plainly indicated by the disturbance of the ground where their tracks met. Two of the tracks were of medium size, one was quite large and evidently made by a grizzly, and the fourth was enormous, being fourteen inches long and nine inches wide.

The last-named track was not made by a grizzly however. There were six toes on the forefoot, and this peculiar deformity was the distinguishing mark of a gigantic cinnamon bear known to hunters as "Six-Toed Pete."

It was almost invariably found, during the long campaign in the wilderness, that tracks over eleven inches in length were made by cinnamon bears, and not by genuine grizzlies, although some hunters declare that the cinnamon is only a variety of grizzly, and that the color is not the mark of a different species. However that may be, the difference between the two varieties is very distinct, and as the object of the expedition was the capture of an indubitable California grizzly, no special effort was made to trap any of the big cinnamons.

The smaller bears soon gave up the contest for the beef and left the field to Pete and the grizzly, who quarreled and fought around it for several nights. At last the grizzly gave Pete a thorough licking and established his own right to the title of monarch of the mountain. The decisive battle occurred one moonlight night and was witnessed from a safe perch in a fork of a tree near the trap.

It was nearly 9 o'clock when the snapping of dry sticks indicated the approach of a heavy animal through the brush, and in a few moments the big grizzly came into sight, walking slowly and sniffing suspiciously. A smart breeze was drawing down the canyon, and the bear, being to the windward, could not smell the man up the tree, but he approached the meat cautiously and seemed in no hurry for his supper. While he was reconnoitering another animal was heard smashing through the thicket, and presently the huge bulk of Six-Toed Pete loomed up in the moonlight at the edge of the opening.

At the approach of the cinnamon the grizzly rose upon his haunches and uttered low, hoarse growls, and when the big fellow appeared within twenty feet of him, he launched himself forward with surprising swiftness and struck Pete a blow on the neck that staggered him. It was like one of Sullivan's rushes in the ring, and the blow of that ponderous paw would have knocked out an ox; but Pete was no slouch of a slugger himself, and he quickly recovered and returned the blow with such good will that had the grizzly's head been in the way it would have ached for a week afterward.

Then the fur began to fly.

It was impossible to follow the movements of the combatants in detail, as they sparred, clinched and rolled about, but in a general way Six-Toed Pete seemed to be trying to make his superior weight tell by rushing at the grizzly and knocking him over, while the latter avoided the direct impact of the cinnamon's great bulk by quick turns and a display of agility that was scarcely credible in so unwieldy looking an animal. Once the cinnamon seized the grizzly by the throat and for a moment hushed the latter's fierce growls by choking off his wind, but the grizzly sat down, threw his arm over Pete's neck, placed his other forepaw upon Pete's nose, sunk his claws in deep, and instantly broke the hold. As they parted, the grizzly made a vicious sweep with his right paw and caught Pete on the side of the head. The blow either destroyed the cinnamon's left eye or tore the flesh around it, so that the blood blinded him on that side, for during the rest of the fight he tried to keep his right side toward the grizzly and seemed unable to avoid blows delivered on his left.

For at least a quarter of an hour the combat raged, without an instant's cessation, both belligerents keeping up a terrific growling, punctuated with occasional howls of pain. Neither could get a fair blow at the other's head. Had the grizzly struck the cinnamon with the full force of his tremendous arm, Pete's skull would have surely been smashed. Pete finally got enough, broke away from the Monarch and fled into the brush, a badly used up bear; and he never came back.

Having won his supper by force of arms, the grizzly was no longer suspicious of the bait, and he ate up the best part of a quarter of beef before he left the battle ground. He soon became accustomed to the trap, and regularly came there for his meals, which were gradually placed nearer the door and finally inside the structure. A piece of meat was tied to the trigger, and one morning the door was found closed, and a great ripping and tearing was heard going on inside. The Monarch was caught at last.

Upon the approach of the men, the grizzly became furious and made the heavy logs tremble and shake in his efforts to get out and resent the indignity that had been placed upon him. Had he concentrated his attack on any one spot and been left to wreak his rage without interruption he would have been out in a few hours, but he was not permitted to work long at any place. Wherever he began work he encountered the end of a heavy stake which was jabbed against his nose and head with all the power of a man's arms.

Day and night from the moment he was found in the trap, the Monarch was watched and guarded, and he kept two men busy all the time. Although his attention was distracted from the trap as much as possible, he found time to gnaw and rip a ten-inch log almost in two, and sometimes he made the bark and splinters fly in a way that was calculated to make a nervous man loathe the job of standing guard over him. For six days the Monarch was so busy trying to break jail that he had no time to fool away in eating. Solitary confinement developed in him a most malicious temper and he flew into a rage whenever food was thrown to him.

But his applications for a writ of habeas corpus were persistently denied by a man with a club, and the Monarch at last cooled down a little and condescended to take a light lunch of raw venison. He was given two days for reflection and meditation, and when he seemed to be in a more reasonable mood, the work of preparing him for a visit to the city was begun.

A running noose was made in a stout chain and put into the trap between two of the logs, and when the bear stepped his forepaw into the noose it was drawn taut and held by four men outside. Despite the strain upon the chain the bear easily threw the noose off with his other paw, letting the men fall backwards in a heap on the ground. Again and again the trick was tried but the noose would not hold.

Then the method of working the chain was changed and the noose let down through the top of the trap, and after many failures it was drawn sharply up round his arm near the shoulder, where it held. Ten hours were consumed in the effort to secure one leg and the Monarch fought furiously every minute of the time, biting the chain, seizing it with his paws and charging about in his prison as though he were crazy. He was utterly reckless of consequences to himself, and he bit the iron so savagely that he splintered his teeth and wholly destroyed his longer tushes.

Having secured one leg, it was comparatively easy to get another chain around his other paw and two ropes around his hind legs, and then he was stretched out, spread-eagle fashion, on the floor of the trap.

The next move was to fasten a heavy chain around his neck in such a way that it could not choke him, and to accomplish this it was necessary to muzzle the Monarch. A stick about eighteen inches long and two inches thick was held under his nose, and he promptly seized it in his jaws. Before he dropped it a stout cord was made fast to one end of the stick, passed over his nose, around the other end of the stick, under his jaw, and then wound around his muzzle and the stick in such a way as to bind his jaws together, a turn back of his head holding the gag firmly in place.

The Monarch was now bound, gagged and utterly helpless, but he never ceased roaring with rage at his captors and struggling to get just one blow at them with his paw. It was an easy matter for a man to get upon his back, put a chain collar around his neck, and fasten the heavy chain with a swivel to the collar. The collar was kept in place by a chain rigged like a martingale and passed under his arms and over his back. A stout rope made fast about his body completed the Monarch's fetters and the gag was then removed from the royal mouth. The King of the mountains was a hopeless prisoner—Gulliver, tied hand and foot by the Lilliputians.

The next morning Monarch was lashed upon a rough sled—a contrivance known to lumbermen as a "go-devil"—to make the journey down the mountain. The first team of horses procured to haul him could not be driven anywhere near the bear. They plunged and snorted and became utterly unmanageable, and finally they broke away and ran home. The next team was but little better, and small progress was made the first day.

At night the Monarch was released from the "go-devil" and secured only by his chains to a large tree. The ropes were removed from his legs, and he was allowed considerable freedom to move about, but a close watch was kept upon him. After several futile efforts to break away, he accepted the situation, stretched himself at the foot of the tree and watched the camp-fire all night.

In the morning the ropes were replaced, after a lively combat, and the bear was again lashed to the sled. Four horses were harnessed to it and the journey was resumed. Men with axes and bars went ahead to make a road, and it was with no small amount of labor that they made it passable. The poor old bear was slammed along over the rocks and through the brush, but he never whimpered at the hardest jolts. With all the care that could be observed, it was impossible to make his ride anything but a series of bumps, slides and capsizes, and the progress was slow. At the steep places men held the sled back with ropes and tried to keep it right side up.

Four days on a "go-devil" is no pleasure excursion, even for a tough grizzly, and when the Monarch was released from his uncomfortable vehicle, at the foot of the mountain, he seemed glad to get a chance to stretch himself and rest. For nearly a week he was left free of all fetters except the chain on his neck and the rope around his body, and he spent his days in slumber and his nights eating and digging a great hole in the ground. Having convinced himself that he could neither break his chain nor bite it in two, he accepted the situation with surly resignation and asked only to be let alone and fed decently.

While the bear was recuperating and becoming reconciled to what couldn't be helped, a cage was being built of Oregon pine lumber with an iron-barred door, and when it was finished he was dragged into it by the heels. As soon as he saw the ropes, Monarch knew that mischief was afoot, and when a man began throwing back into the hole the dirt that he had dug out, he mounted the heap and silently but strenuously began to dig for himself a new hole. He worked twice as fast as two men with shovels, and in his efforts to escape he only assisted in filling up the old hole.

For some time he baffled all attempts to get ropes on his forepaws, having learned the trick of throwing them off and seizing the loops with his teeth, but he was soon secured and stretched out on his back. The Monarch roared his remonstrances and did his best to get even for the outrages that had been done to his rights and his feelings, but the ropes were tough and he could not get a chance to use his enormous strength. He was dragged on his back into the cage, the door was dropped and the ropes were removed, but the chain remained around his neck and that was made fast to the bars. As soon as he found himself shut up in a box the angry and insulted bear ceased roaring and in a short time he philosophically stretched himself on the floor and wondered what would happen next.

The next thing that happened to him was the standing of his cage on end, but that did not appear to disturb him. A wagon was backed up, and the cage was tilted down again and placed upon the wagon, which was then hauled down the canyon and along the river bed to a little water station on the Southern Pacific Railroad, where the cage was put upon a stock car. The car was provisioned with a quarter of beef, and a lot of watermelons, and attached to a freight train, then men who had helped to bring the bear out of the mountains waved their hats, and the Monarch caught a last glimpse of his native hills as the train whirled him and the correspondent northward.

It must have been a very strange, perhaps terrifying, thing to the wild grizzly to be jolted along for two days on a rattling, bumping, lurching freight train, with the shrieking of steam whistles and the ringing of bells, but he endured it all heroically and gave no sign of fear. He ate well when food was given him, taking meat from his captor's hands through the bars, and slept soundly when he was tired. He seemed to know and yield a sort of obedience to the correspondent, but resented with menacing growls the impertinent curiosity of strangers who came to look at him through the bars.

In every crowd that, came to see him there was at least one fool afflicted with a desire to poke the bear with a stick, and constant vigilance was necessary to prevent such witless persons from enraging him. At Mojave, when the correspondent went to the car, he found a dozen idlers inside, and one inspired lunatic was stirring up the Monarch, who was rapidly losing his temper. The cage would not have held him five minutes had he once tackled the bars in a rage, and it was only the moral influence of the chain around his neck that kept him quiet. When the correspondent sprang into the car, the grizzly's eyes were green with anger, and in a moment more there would have been the liveliest kind of a circus on that freight train. Hustling the crowd out with unceremonious haste—incidentally throwing a few maledictions at the man with the stick—the correspondent drove the Monarch back from the bars, and ordered him to lie down, and for the next half hour rode in the car with him and talked him into a peaceable frame of mind.

From the freight depot on Townsend Street the cage was hauled on a truck to Woodward's Gardens, and under the directions of Louis Ohnimus, superintendent of the gardens, the Monarch was transferred to more comfortable quarters. His cage was backed up to one of the permanent cages, both doors were opened, and he was invited to move, but he refused to budge until his chain was passed around the bars and hauled by four stout men. The grizzly resisted for a few minutes, but suddenly decided to change his quarters and went with a rush and a roar, wheeling about and striking savagely through the bars at the men. But Mr. Ohnimus had expected just such a performance and taken such precautions that nobody was hurt and no damage done.

The Monarch had shown himself a brave fighter and an animal of unusual courage in every way. He had endured the roughest kind of a journey without weakening, and compelled respect and admiration from the moment of his capture. But when the strain and excitement were over, and he was left to himself, the effects became apparent, and for two or three days he was a sick bear. He had a fever and would not eat for a time, but Mr. Ohnimus took charge of him, doctored him with medicines good for the ills of bear flesh, and soon tempted back his appetite with rabbits and pigeons.

Soon the Monarch was sufficiently convalescent to rip the sheet iron from the side of his cage and break a hole through into the hyena's quarters. By night he was on his muscle in great shape, and Superintendent Ohnimus sent for the correspondent to sit up with him all night and help keep the half-ton grizzly from tearing things to pieces. By watching the old fellow and talking to him now and then they managed to distract his attention from mischief most of the time, but he got in considerable work and rolled up several sheets of iron as though they were paper.

It was evident that no ordinary cage would hold him, and men were at once employed to line one of the compartments with heavy iron of the toughest quality and to strengthen it with bars and angle iron. This made a perfectly secure place of confinement. A watch was kept on the Monarch by the garden keepers during the day, and by the superintendent and the correspondent every night, until the work was finished and the Monarch transferred.

The grizzly is now safely housed in the first apartment of the line of cages, and under the watchful care of Mr. Ohnimus will soon recover his lost flesh and energy and again be the magnificent animal that he was when he was the undisputed monarch of the Sierra Madre.

LATEST BULLETIN.
Monarch a True Grizzly.

"Monarch," the Examiner's big grizzly, received many visitors yesterday, but, having been up all night trying the strength of his new house, he declined to stand up, and paid but little attention to the crowd. His chain had been fastened to the bars of his cage with three half hitches and a knot, and the knot was held in place by a piece of wire. During the night he removed the wire, untied all the knots and half hitches and hauled the chain inside, where nobody could meddle with it. Having the chain all to himself, Monarch was indifferent to his visitors and lazily stretched himself on his back, with one arm thrown back over his head.

He had a good appetite yesterday and got away with a leg of lamb and a lot of bread and apples. He ate a little too heartily and had the symptoms of fever. Today he will not get so much food. The best time to see him is when he eats, because he lies down all other times of the day. He has breakfast at 10 a. m., lunch at 1 p. m. and dinner at 3 p. m.

Monarch still looks travelworn and thin, but he is brightening up, and when the abrasions of the skin, made by ropes and chains, are healed up and his hair grown again on the bare spots he will be more presentable. His broken teeth trouble him some and it will be some time before he will feel as well as he did before he was caught.

Several artists went to Woodward's Gardens today to sketch and photograph the bear, but he refused to pose, so they did not get the best results. It would be unwise to stir him up and excite him at present, and unless the artists can catch him at his meals they will have to wait a little while for a chance to study the grizzly under favorable conditions.

Sculptor Rupert Schmidt has made an excellent model in clay of Monarch, which will be a valuable assistance in designs requiring the introduction of the California emblem.

Mr. Schmidt said:

"I am very glad to have the opportunity to study the real grizzly, and I find him very different from the models generally accepted. I have modeled many bears, but never one like this. You see in this design some figures of bears (showing a wax model of decorative capitals). These were intended to be grizzlies, but you see they have the Roman nose, which is characteristic of the black bear. No other bear that I ever saw had the broad forehead and strong, straight nose of the grizzly. He has a magnificent head, and I think all artists will be glad of a chance to study him. I have inquired for grizzlies in zoölogical gardens all over the world, but never found one before."

Monarch has a big, intelligent-looking head and a kindly eye, and is not disposed to quarrel with visitors, but he objects to any meddling with his chain, and will not submit to any insults. It was necessary yesterday to keep a watchman between the cage and the crowd to prevent people from throwing things at the bear and stirring him up. Monarch is getting along very well and taking his troubles quite philosophically; but he has had a rough experience, is worn out with fighting and worry, is sore in body and spirit and needs rest. It is a difficult thing to keep alive in captivity a wild bear of his age, and undue excitement might throw him into a fatal fever. If Superintendent Ohnimus succeeds in his efforts to cure the Monarch of his bruises and put him into good condition, he will deserve great credit, and the visitors are requested not to make the task more difficult by worrying the captive. No other zoölogical garden in the world has a California grizzly, and it would be a great loss to the menagerie to be established in the Park if the Monarch should die.

It is not surprising that many people cannot tell a grizzly bear, even when they see one, as many zoölogists even differ widely in regard to the characteristics of the king of bears. It is astonishing how little is really known in regard to the grizzly bear. Many text-books contain only a general notice of the great animal, while those naturalists who have written descriptions of him do by no means agree. This is due to their lack of specimens. The grizzly is so powerful and unyielding a beast that but few have been captured alive. There have not been individuals enough of the species studied to admit of their being fully generalized. Different naturalists described the grizzly from the single specimen that came within their notice, and hence their various descriptions are far apart. It is a fact that hardly two of the animals taken are exactly alike in color or habits.

In order to definitely settle the question, Prof. Walter E. Bryant, of the Academy of Sciences, was yesterday induced to visit the bear. He has made the mammals of the Pacific Coast his study for years, and probably knows more than anyone else about California bears.

He examined Monarch very carefully, noted his every point, and then examined just as carefully the other bears at the gardens.

When he had completed his investigation and stood once more before Monarch's cage, he was asked:

"Well, what is he?"

"He is a true grizzly bear," answered Professor Bryant, and he added, "a mighty big one, too.

"I never before saw one of the animals with as dark a coat as his," he continued; "but that is nothing. The bear is a true grizzly, and has all the characteristics of one. As far as his color is concerned, grizzlies are of all colors; there is almost as much variety in that regard among bears as among dogs."

"How do you know it is a grizzly?" was asked.

"Well, in the first place, the claws on his forefeet are longer and stronger than those of any other species. Then his head is larger than that of other bears, and his muzzle is longer and heavier. Another and more distinguishing feature is the height of his shoulders. Just back of his neck is the tallest point. From there his back slopes down towards his haunches. The black bear, on the other hand, has low shoulders, and is tallest at a point rather back of the middle of the body. There are numerous other means of distinguishing this bear. His teeth are very much larger and stronger than those of the others, and the entire structure of the skull is peculiar to the grizzly. He has neither the short muzzle of the European bear such as you see in the pit, nor the rounded muzzle of the black bear. There are, of course, many minor points that only a naturalist would observe, but it is sufficient to say that he lacks none of the essential qualities of the grizzly bear, and has none of those of the other varieties.

"His coat is almost black, to be sure, but it is very different from the glossy black of his neighbor. If you observe the grizzly's hair, you will see that a great deal of it is a rusty brown and in certain lights seems to be very far from black. This variation in the color of the hair is a peculiar characteristic of the grizzly. That lanky mane is another. His legs, you observe, are darker than his body. This is another characteristic of the California grizzly.

"This animal is thin now, doubtless from the hard time he had while he was being brought here. When he gets fat his hair will have a very different appearance. It will be interesting to watch him when he sheds his hair. The coat that comes after may be altogether of another color. That grizzly, I should say, is comparatively a young bear, and when he gets older the gray that originally gave him his name will very likely be pronounced."

THE END.

SCIENTIFIC CLASSIFICATION OF BEARS.

Edited by PIERRE N. BERINGER.

I.

THE LOUISIANA SPECTACLED BEAR.
Tremarctos Ornatus.

Some of our scientists have very carefully divided the *Genus Ursus* into *twelve* species. While I will admit that these gentlemen are conscientious and that they are thorough in their researches, I wish to point to the fact that they have entirely overlooked three or four species found on the Pacific Coast.

Many writers have completely ignored the spectacled bear of Louisiana. Is he the representative of another genus? Does he belong to the *Genus Helarctos* (*helios*, the "sun," and *arctos*, "bear") credited by the majority of writers with basking in the sun, or because of the peculiar markings of his chest, representing a sunburst? He resembles the *Helarctos Malayanus* of the Malayan archipelago or the *Bruang* of Java. Or is he the Sloth Bear, *Prochilous* (or *Melursus*) *labiatus*? This bear has been carefully classified as a separate genus found from the Ganges to Ceylon. His description fits rather loosely the so-called sloth of Louisiana. Possibly the Louisiana specimen is of the *Genus Tremarctos*, of which the learned people tell us there is but a solitary species carefully isolated in the Andes of Chile and Peru. I shall call the Louisiana specimen by the name given him by our poet, the Spectacled Bear, *Tremarctos Ornatus*, and the professors who have entirely overlooked his existence may classify him later when they find time. At one time the Honey Bear was classified as a "Bradipus," or sloth, because of its liability to lose its incisors. It was therefore set down as one of the *Edentata*. It has also been styled the Jungle Bear, the Lipped Bear, and names as various as the investigators' fancy. The *Tremarctos Ornatus* of Louisiana, or spectacled bear, is not a sloth. He does not belong to the *Edentata*, neither is he lazy. He is essentially the clown of all bears, a very intelligent animal, and in many cases the intellectual superior of his keeper. He is active to a degree, and will perform the queerest antics for the amusement of the onlooker. He is quaintly conscious of his mirth-provoking powers, much as a child playing "smarty." He will quickly climb an inclined log or tree, and then slide down either in an upright position, clasping the log with the knees, or he will slide "down the banister" as a child might. I have seen the merry fellow grab his tail in his mouth and roll over and over until dizzy.

His snout is almost hairless, narrow and proboscis like, and the nostrils and lips are mobile. He shapes these almost into a pipe, through which his long tongue is shot out, drawing things in or sucking them up. It has claws of a bluish gray that are longer than those of any other of the Ursidae. The hair is very long, of a deep brown black. There is a sunburst upon the chest of a white or fulvous hue. The ears are small and scarcely distinguishable, owing to the shaggy mane. The fur is rather coarse and very long.

It lives mainly upon honey and vegetables and sugar cane. In captivity it will very gratefully subsist upon oatmeal and occasional sweets. The animal is easily tamed, and will become attached to its keeper, giving an exhibition of exuberant joy at his approach. It is a jolly good fellow, and shows a marked preference for liquors, refusing all others when it may have champagne.

It will sit on its hind legs and make faces at the onlooker, waving its arms in the most grotesque fashion, while it rolls its body from side to side. This is one of the characteristics that has impressed the negro with the sacredness of this "Voodoo Bear."

II.
THE GRIZZLY.
Ursus Horribilis or Ferox.

This is the great grizzly of California, whose habits have been described by many writers. It is a shy animal, not nearly as ferocious as has been claimed. "It will always run away if it can," says General Dodge, "and never attacks unless it is cornered or wounded." Johnson says "the grizzly is the king of all our animals, and can destroy by blows from his paws the powerful bison of the plains; wolves will not even touch the carcass of the dreaded monster, and, it is said, stand in such awe that they refrain from molesting deer that he has slain. Horses also require careful training before they can be taught to allow its hide to be placed upon their backs."

In the beautiful legend of the Good Poet the grizzly is the forefather of the Indian, and the Indian gives many proofs to show his descent from the grizzly and the Spirit of the Mountain. I want to add a curious fact: The grizzly is the only one of the Ursidae that moves his toes and fingers independently of one another just like a *man*. Also the bear walks with his foot full upon the ground. In further proof the grizzly, when young, and all other bears, except one, descend a tree backward and head up, as a man would. The clown bear, or spectacled bear, will sometimes descend head down and enjoy a good laugh over it. At least he seems to laugh. After the grizzly has attained bulk and weight with age, he cannot climb trees, as his claws are not strong enough to sustain his weight.

A short time after "Monarch," the large grizzly, arrived in San Francisco, my model, a very considerate young person, who loved all animals, came to the studio one day with the story that she had made friends with the great beast. It was about the time when "Monarch" was being starved. He had been removed from the pit to the cage. With very little forethought the cage was built without a cover, and "Monarch" was found one night making an attempt to escape. He was prodded back with red-hot irons. It was not possible to work about the cage, and "Monarch" must be confined in smaller quarters. A very small cage was dropped into the enclosure; this had a slide door and was to serve as a trap. I believe the grizzly is the quickest of all animals. Six times a live chicken was fastened in the small cage, and six times "Monarch's" long arm had literally "swiped" that fowl. So quick was he that the slide fell only as he was already safely crunching its bones. At the seventh attempt he was a little slow and was caught. After that the iron workers placed the roof in position. The trapping of the monster took six days, and "Monarch" received only the food he managed to get from the trap, and that which my tender-hearted model was feeding him (apples and candy) surreptitiously. As this was against the orders of the keeper, the young woman could feed the bear only at irregular intervals. She continued her kindnesses to him after he had been again given the freedom of the larger cage. Then she went away from the city. She was gone for two years. She married and assumed the rotund proportions of a staid matron, and when next I saw her I joked her about this, saying that she was nearly as fat as her old friend "Monarch."

At this she was indignant. "Indeed," she said, "animals are less forgetful than man, and 'Monarch' undoubtedly will remember me, even if I am not the slim artist's model I once was." I told her "Monarch" was far too much like a man, and that he was now satisfied to look upon the world as well lost, and that short of his dinner there was little that could move him from a comfortable position upon his back, his toes in the air, apparently content, and like a philosopher, wondering why the human displays so much curiosity. "I'll bet he won't stir," I said. The upshot of this conversation was that we found ourselves just outside the railing gazing at his lazy majesty. He rolled his head slowly from side to side, eyeing each newcomer with his bead-like eyes. Suddenly the lady in the case said, "Oh, you dear old darling!" "Monarch" seemed electrified; he rose as quickly as possible—certainly he had grown fat—and then he rushed to the side of the cage. He was not satisfied with looking at her from his ordinary standpoint, but rose upon his feet, extending himself his entire height, that he might better look upon the friend of times of trouble. She held up an apple. "Monarch" dropped to his feet, placed his snout as far out as the bars allowed, and opened his immense jaws. She threw the apple, and the bear sat himself down contentedly to chew it. I firmly believe that young woman could have walked into the cage with an apron full of apples and escaped without injury. "Monarch" remembered his friend.

III.

THE POLAR BEAR.

Thalassarctos Maritimus.

Much uncertainty prevails respecting the generic classification of the bears. Wallace has divided them into five genera or subgenera, and fifteen species. Wood gives eighteen, and Gray says twelve. The appearance of the bear at different seasons has led to much error in classification. The practical mountaineer will tell you of some three or four species in California that have been given notice of as the young of another species, or that have never been mentioned by the learned gentlemen who usually study bear life in the seclusion of a library or with the help of a strong field telescope. A glance at the teeth of the bear will tell you that they incline rather to the vegetable diet. Their ferocity is almost always exaggerated. Their courage is desperate in self-defense, but it is seldom that they become the aggressor. The brain of the bear is very highly developed, and they soon learn all kinds of accomplishments. The lion is an uncouth boor in comparison.

The Polar Bear, *Thalassarctos Maritimus*, is the only representative of the genus. He is an almost wholly carnivorous animal, his food consisting of fishes and seals, which he skillfully captures. He can swim better than any other bear, and has been known to swim a strait forty miles wide. The fur is silver white tinged with yellow. This color is variable in specimens, and according to the seasons. The head is much smaller than that of the grizzly or black bear, and is ferret-like, with a decided downward curve to the nose. The nose does not possess the flexibility of that of the rest of the bear family, although the polar bear has the higher development of the sense of smell. Johnson says that the flesh is good to eat, but other writers do not agree with him. Kane was poisoned by eating of the liver. In speaking of a capture De Vere wrote as follows: "We dressed her liver and ate it, which in taste liked us well, but it made us all sick * * * for all their skins came off, from the foot to the head, but they recovered again, for which we gave God hearty thanks." Hall says that the Eskimos of Cumberland Sound likewise believe the liver to be poisonous, even for the dogs. Ross says all who partook of the meat suffered from severe headaches, and later the skin peeled from the body. Greely says his party largely lived upon the meat, and that it was coarse, tough, the fat having a decidedly rank flavor.

I believe that the physiognomist may follow the characteristics of an animal by his facial expression, and that with the aid of a knowledge of the cranial development he can gauge the mental caliber of the beast. Following this system and adding to it the testimony of credible explorers, it is quickly shown that the polar bear is treacherous and intractable. While he is not the wise animal the grizzly is, he is more cunning and is certainly not a coward.

There are times when he is not content with being let alone, but will take the aggressive. Greely writes: "Doctor Copeland was surprised only fifty yards from the ship by a bear which broke from a barrier of ice hummocks, galloped up to within five paces, reared up and struck him down with both forepaws. Copeland had no time to load his gun, but as the animal caught his clothes, he swung the butt across his snout. This and the noise of approaching comrades put the bear to flight, and he started off with the swinging gallop peculiar with him."

The mother bear and cubs display a great fondness for one another. Koldeway says: "No sooner did the young ones perceive the hunter than they galloped toward their mother, who in two strides turned and stood by them, with such rage expressed in all her actions that we knew we must be careful. Finding, however, that they were unhurt, she seemed to think only of bringing them to a place of safety."

Some authorities have it that only the she bear hibernates and that the male continues in the active exercise of all his faculties. Ross weighed a polar bear which tipped the scales at 1,131 pounds; Lyon saw one which weighed 1,600 pounds; Dr. Neale tells of one measuring eleven feet exclusive of the tail. Senator Wm. P. Frye has the skin of one, presented to him by an explorer, which measures nine feet seven inches exclusive of the tail of two inches. Its girth around the body just back of the forelegs is ten feet.

IV.

THE CINNAMON BEAR.

Ursus Cinnamoneous.

The Cinnamon Bear has been called a variety of black bear. I am inclined to believe it a separate genus. The head has many points of difference. It is wider. The eyes are set deeper, and closer together. There is a better breadth of brain. The feet are smaller. The fur is rather longer than that of the black bear and much softer. The color is dark chestnut, and as the bear ages there is an occasional gray hair. The cinnamon is more dignified than the black bear, and he also remembers an injury longer. A baby cinnamon was captured by a friend of mine and brought to the city. A chain was placed about its neck, and this was attached to a peg that was hammered in the ground. As soon as I heard of the coming of his bearship I hurried over and made his acquaintance. He ate a quart of milk soaked into as much bread as it would hold, and enjoyed it greatly. He chewed on my finger every time I dipped it into sugared water without biting.

I left him fast asleep. When I returned in the afternoon he was walking from side to side, shaking his head, and howling most dolefully. The cry was much like that of a child, only louder and more disagreeable. He was hungry. I went to him and I said, "Stop it." At this he howled so it made my head ache. I picked him up, and with the aid of a shingle, I gave him a spanking, just as you would a bad boy. This stopped his howling, and then his master came and fed him.

After this spanking it was very evident that he did not care for my acquaintance. He persistently refused to recognize me. As I approached him his ears would go back, and his fur would rise. He had decided to cut my acquaintance.

Some days after, I was watching a tennis game in the next yard, standing with my back to baby bruin. He couldn't overlook the opportunity to get even, and, watching his chance, he fastened his teeth in the calf of my leg.

V.

THE BLACK BEAR OF CALIFORNIA.

Ursus Californiensis.

This bear we will label for convenience *Ursus Californiensis*, because the title of *Ursus Americanus* has dignified the small black bear of the Eastern states. There are, however, three species of the black bear in California that are known, and there may be more. The large black bear of California reaches very large proportions. I have seen some that might weigh from 800 to 1,000 pounds. It is hunted for its fur, which is uniform in color, and for its flesh, which is quite good, either smoked or fresh. This animal will never seek an encounter with man. I remember my original introduction to a bear of this species. It was in the state of Washington. Owing to ill health I had been staying at what is known there as a ranch. A ranch in the western Washington forests generally consists of a shake hut or log house, and a promise by the "rancher" that he will soon clear enough ground to raise *something*. Generally this vague something is a mortgage. This particular rancher had a cow, and this cow often strayed away into the timber and had to be looked after when milking time came.

One day, in the exuberance of new found health, I was taking the greatest of pleasure in chasing that cow toward the "shed" road to the ranch. I was feeling especially good, and I was jumping over fallen trees, making short cuts and throwing broken branches and an occasional stone at the old Jersey.

Suddenly I stopped before an extra high log, and gathering myself together, I jumped high over it. I landed upon the upturned belly of an old she bear. There was a sound like the escape of gas from a rubber bag. I passed the cow like a streak of lightning. When I had run a considerable distance I turned my head and saw the bear running in the opposite direction. I did not stop, however, and I got to the ranch nearly an hour before the old cow.

In the shingle mills of the North the Norwegian hands have the same veneration for the bear as the Indian. They always speak of him not as a bear, but as "the old man with the fur coat on."

Large Black Bear.

VI.

QUAINT INDIAN LORE

IN REGARD TO THE MYSTICAL POWER OF THE BEAR AS A GREAT MEDICINE.

This is a legend of the Ojibwa Indians as told by Sikassige, the officiating priest of the Ojibwas at White Earth, Minnesota:

In the beginning were created two men and two women. They had no power of thought or reason. Then the Almighty took them into his hands that they might multiply, and he made them reasonable beings. He paired them, and from this sprung the Indians.

Then when there were people the Great Spirit placed them upon the earth; but he soon observed that they were subject to sickness, misery and death. Then the Manitou called upon the Sun Spirit (the Bear) and asked him to instruct the people in the Sacred Medicine. The Sun Spirit, in the form of a little boy, went to the earth and was adopted by a woman who had a little boy of her own.

This family went away in the autumn to hunt, and during the winter the woman's son died. The parents were much distressed and decided to return to the village and bury the body there. So they made preparation to return, and as they traveled along they would each evening erect poles upon which the body was placed, to prevent the wild beasts from devouring it. When the dead boy was thus hanging upon the poles, the adopted child, the Bear Spirit, or Sun child, would play about the camp and amuse himself, and finally told his adopted father he pitied him and his mother for their sorrow.

The adopted son said he could bring his dead brother to life, whereupon the parents expressed great surprise and desired to know how that could be accomplished.

The adopted boy then had the party hasten to the village, when he said: "Get the woman to the wigwam of bark, put the dead body in a covering of birch bark, and place the body on the ground in the middle of the wigwam." On the next morning, when this had been done, the family and friends went into the lodge and seated themselves around the corpse.

Then they saw, through the doorway, the approach of a bear, which gradually came toward the wigwam, entered it, and placed itself before the dead body, and said "Hu, hu, hu," when he passed around toward the left side, with a trembling motion, and as he did so, the body began quivering, which increased as the bear continued, until he had passed around four times, when the body came to life and stood up. Then the bear called to the father, who was sitting in the distant right hand corner of the wigwam, and said:

"My father is not an Indian. You are a spirit son. Insomuch my fellow spirit now as you are. My father now tobacco you shall put. He speaks of only once to be able to do it. Why he shall live here now that he scarcely lives; my fellow spirit I shall now go home."

The little bear boy was the one who did this. He then remained among the Indians and taught them the mysteries of the Grand Medicine, which would assist them to live. He also said his spirit could bring a body to life but once, and he would now return to the sun, from which they would feel his influence.

This is called "Kwi-wi-senswed-di-tshi-ge-wi-nip"—"Little boy, his work."

VII.

CURIOUS FACTS ABOUT THE BEAR.

With the different seasons the bear presents a varied appearance. There are times when you would scarce recognize the same animal. In the autumn of the year the bear takes on fat in preparation for hibernating. At this time the fur is glossy and long, and in the grizzly almost a seal brown.

A curious phenomenon now takes place in the animal's digestive organs, which gives it the capability of remaining the entire winter in a state of lethargy, without food and yet without losing condition. As the stomach is no longer furnished with food, it soon becomes quite empty, and, together with the intestines, is contracted into a very small space. No food can now pass through the system, for an obstruction, a mechanical one—technically called the "tappen"—blocks the entrance to the passage and remains in this position until spring. The "tappen" is composed almost entirely of pine leaves and the various substances which the bear scratches out of the ants' nests or the hives of bees. During the season of hibernation, the bear gains a new skin on its feet. It will remain in its den until about the middle of April or the beginning of May, and will emerge almost as fat as when it entered, unless it has lost the "tappen" too soon.

It will now be seen that the fur has undergone a change. With the grizzly it has the real grizzly hue; with the brown or black bear it has a dead look. This is the hungry season for the bear, and until fall, when the berries are ripe and the salmon run in the streams, his bearship has a hard time of it. By the end of July and until the middle of August the fur undergoes a further change. The old coat is hanging upon him in shreds, he is much emaciated, and there is a hungry look in his eye. His ears appear abnormally large, and his paws seem enormous. When the berries are ripe and there are fish in the streams, the preparation for winter begins, the fur is sleek and greasy-looking again. Mr. Bear is fat and contented and ready to go into his long sleep.

When he awakes one of the first things he does is to suck his feet. This is done because the skin is new and tender.

In the picture illustrating the fight between the bear and the boy upon the log, the bear is shown as he appears during the emaciated season, a caricature of himself when well fed. The bear in captivity receives his food at regular intervals and in large quantities, and he loses many of the marked characteristics of the bear in his wild or untamed state. There is just as much difference between a society leader and a man who lives close to nature.
